Data Energy Matter

Architecture Climate Changes

David Serero
Serero Architects

Table of Contents

MATTER

Foreword

Data Energy Matter Architecture Climate Changes

David Serero

The exploration of architectural practice over twenty years is sufficient time to witness the crystallization of fundamental ideas and approaches in architectural design. The projects selected here reveal a recurrence of key figures and concepts that unfold and dissipate throughout the work.

The projects challenge the notion of architecture being viewed as a permanent structure, and instead, they advocate for a transitory and ephemeral approach with greater spatial versatility. It is an "open architecture", which encourages sharing, community building and interconnection, like a mycelium network.

Our in-depth observation of natural processes has taught us that time is the best ally of adaptation, transformation and co-creation. Trees, often used as metaphors and symbols in our work embody the optimization and redundancy necessary for adaptability and contextual transformation. The influence of architects such as Alvar Aalto and Frank Lloyd Wright emphasize the way in which nature inspires architectural projects and experimentation.

Over the years, we have learned that buildings are intrinsically linked to the concept of energy, not only in terms of consumption but also in terms of production and sharing. This energy paradigm will redefine architecture. The buildings of tomorrow will not only save energy, they will produce and share it, thus helping to address the environmental crisis.

We view buildings in terms of energy and resilience. For us, architecture must minimize its impact on the environment and integrate with nature. Like a tree, architecture must unfold in its context, breathe with its environment, and forge new links with its surroundings. Our thinking is inspired by models found in nature. It does not aim to create new spaces but to deploy models of cultural, social, and spatial origins.

Three central sources emerge from the projects presented in this book. Rather than the three principles of architecture, its sources are to be considered as the triggers of architecture rather than an objective of architecture.

Data

Digital data and programming codes constitute a new field of research and knowledge that architects must acquire and integrate into their work. The architect-researcher thus sets up systems for generating and analyzing data and develops an informed design method of the Design by Data type. When the architect interacts directly with the analysis tools, it allows him to better understand the impact of each of his choices on the project environment and its urban and territorial context. This method of architectural design complements the methods used by architects: parallel design from one scale of the project to another, refinement of the overall meaning of the project down to its finest expression in the architectural details, creation of prototypes, and prefiguration of the assembly of the project's materials, which are proven methods for analyzing, understanding and designing a space.

Parametric design, collaborative workflows, augmented design, and AI-driven processes have transformed our projects. The mathematics of architecture has evolved from dimensions and proportions to technical processes driven by data ecosystems. These advances feed the design process, but they also challenge the traditional role of the architect. Understanding the logic of machines, neural networks, and algorithms will be essential to visualize new modes of collaboration between humans and machines.

In this new data era, architects risk losing their leadership in architectural creation as the process is increasingly distributed between people and tools. The advent of AI generating text-images, text-models, and imminent speech-architecture calls into question one of the last bastions of the architect's role: ideation. Will architects become mere assistants to machines? We firmly believe that mastering the logic of machines, neural networks and algorithms is crucial for controlling and improving the collaboration between human creativity and technological innovation.

Energy

Our philosophy defines the energy issue as inseparable from the architectural issue. Energy and architecture are intrinsically linked, forming a common objective that shapes both the architectural design of the project and its environmental approach.

Architecture is also a dynamic environment that generates interior ambiance, regulates solar radiation and ventilation, natural ventilation, and limits energy requirements. Almost all our projects incorporate bioclimatic devices as fundamental elements, shaping the very image of our designs.

Our research has largely focused on the intelligence and adaptability of the building envelope. Recently, as part of a global research initiative, we have been studying natural ventilation and air circulation around buildings. Using advanced simulation software, we have developed methods to facilitate design decisions using tree dendrograms, enabling virtual experiments with light, wind behavior, and heat radiation.

It anticipates an era of energy symbiosis, where buildings are not mere consumers but vital nodes in an energy-sharing network. Micro-energy production has become increasingly important, with devices gradually being integrated into the skin of buildings. This development is facilitated by the emergence of smart grids or the energy internet, where buildings, neighborhoods, and cities share and exchange energy, transforming each entity into an active energy producer and distributor.

Our vision includes fully integrated wind and solar energy systems combined with adaptive façades that passively and continuously adjust to local conditions. This transformative approach ensures that architecture is not only reactive but also actively participates in the creation of sustainable and energy-free environments.

Matter

Matter is a major component of the project design process along three lines: Matter-history, Matter-experience or Matter-resources. It is deployed to create a dialogue with the history of the place in some projects, or to create a physical or conceptual experience of the matter in others, or even as an alternative construction method using reclaimed matter in a militant stance to preserve natural resources.

In the context of old towns and heritage buildings, materiality is a force for dialogue, enabling a link to be forged across the ages between ancient and contemporary architecture.

The experience of matter in the project places the material before space or volume. It is a generative force of architecture.

With resource matter, we deeply root our structures in their environment, using local materials to activate a cycle of sustainable materials with recovered, found, or discarded resources.

This approach promotes a radical change: buildings are no longer mere constructions but resource banks of materials, echoing the practices of our predecessors who reused stone and wood from existing structures. With the increasing scarcity of building materials and our unwavering commitment to reducing the carbon footprint of buildings, we are confronted with a staggering truth: architects will have to build less and probably focus on reimagining existing structures or already urbanized sites instead of creating new ones.

For us, architecture is an "informed matter". It is the catalyst of the emerging cycle of materials. It embodies the digital revolution, revolutionizing creative work in all dimensions of the project.

This analysis of our practice of architecture is guided by data, energy, and material. These projects, which combine an experimental method, thus contribute to the definition of research in architecture. If the architect develops, in certain cases, new knowledge about the project, its design, or its construction, does he not generate an approach that constitutes a contribution to the epistemology of architecture or to design methods?

Chapter 1

DATA ENERGY MATTER

Data
From Natural Processes to Artificial Neural Networks

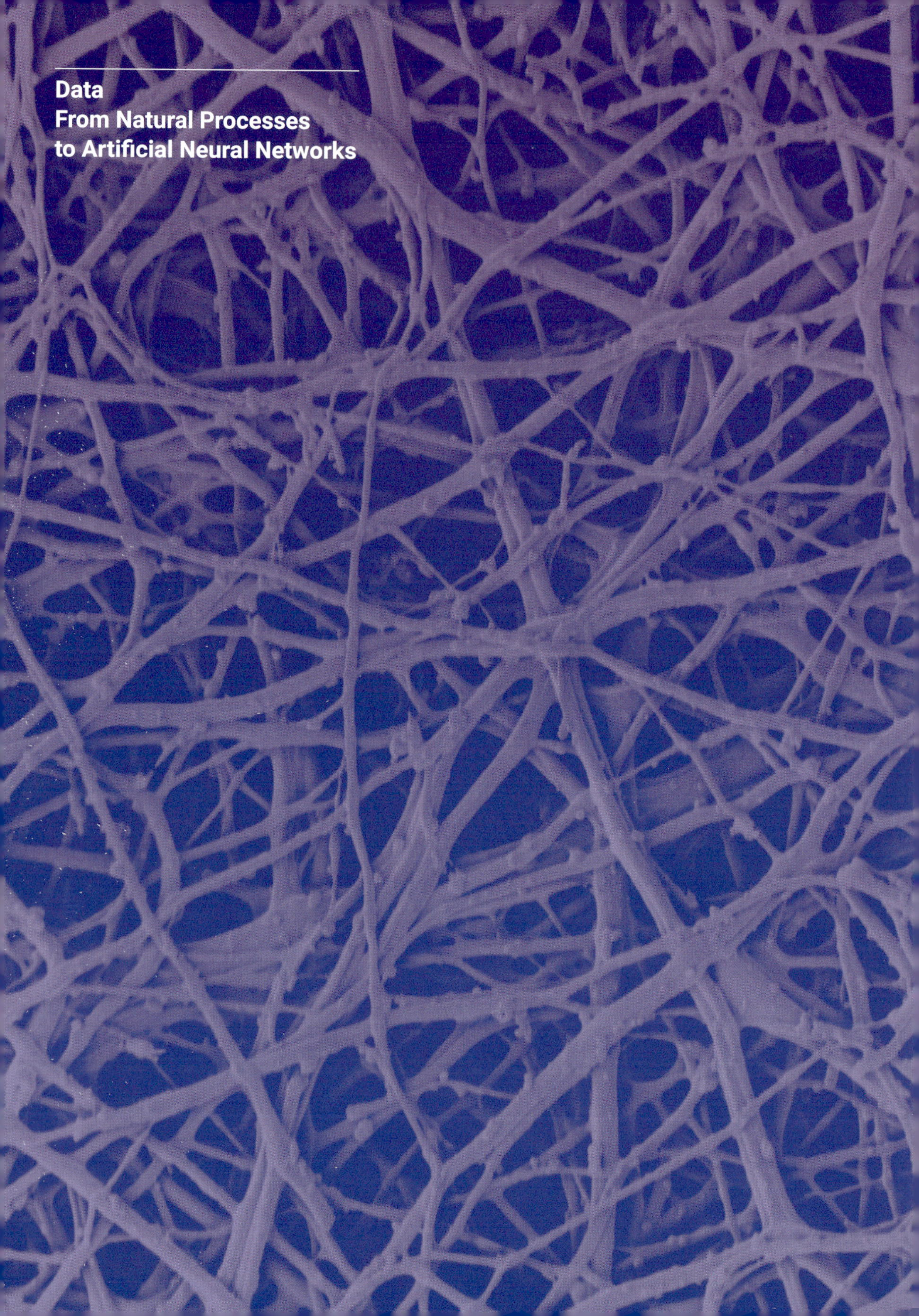

The use of data in project design opens up a new relationship between mathematics and the project design model. While the discussion on the dimensions and proportions of the building was central in the old treatises on architecture, the advent of Cartesian space, the digital model, parametric design, augmented design by large artificial intelligence language models have opened up new conceptual and technical potentials for architecture.

While these advances nurture and irrigate the process of designing new methods, they also call into question the traditional role of the architect. They encourage a rethinking of the architect's mission by intensifying the collaboration between man and machine based on the mastery of machine logic, neural networks, and algorithms. The projects presented here address a number of reflections on these issues.

The Ghost Track project for the 2002 Venice Biennale, created with A. Descombes and A. Regnault, as part of the DZo group, presciently addressed the resistance and inability of digital models and 3D technology to perform fine operations on materials, such as the principles of weaving materials to form a structure. The Ghostrack installation at the French pavilion was designed on the basis of a digital model and made by hand, unlike other works of that period, which were based on the use of 3D printing technologies to produce geometries designed in the conventional way.

The Saint Cyprien auditorium projects, the variable-geometry dome in Rome, the Meudon-la-Forêt cultural center, and the Eiffel

DNA are explorations of generative geometry technologies for the project. A parametric model is coded and then used as a process for generating forms for the project that irrigate all scales of the design. In this model, these codes are often inspired by natural generative processes—themselves derived from algorithmic logic (Wolfram, 2002)—which are diverted or appropriated within the framework of the project.

For the Saint Cyprien auditorium, branching and tree-bearing models are coded with the aim of generating a complex of openwork roof and ceiling to create luminous undergrowth that encompasses all the spaces of the building. A similar model was then coded to generate the parametric foliage for the shape of the oblong bays of the Meudon-la-Forêt cultural center.

On the other hand, the ADN Eiffel project completely digitized the structural profile of the Eiffel Tower in a parametric model to create, by extension of this profile, an extension of the form, like a budding of this century-old structure, in order to propose a temporary platform for the Eiffel Tower.

In the work carried out as part of our residency at the Villa Medici in Rome, the project for a dome with variable geometry is the result of research on the displacement of acoustic waves in the moldings of the classical interior spaces of a large reception hall. This propagation model made it possible to define the profile of a variable acoustic shell surface, making it possible to adapt the acoustic response of the space to the types of performance: singing, chamber music, amplified music, and conferences. These are all usage scenarios where the adaptability of the form makes it possible to modify the acoustic experience of an event.

In the case of the Modulations for a temporary garden in Quebec project, the fern's model of plant growth was reused to create a stainless steel screen that is both a plant twin and a winding guide for visitors to the garden.

For the Learning Center in Issy-les-Moulineaux, the restructuring of the building takes a fractal model, that of the Menger-Sierpinski sponge, as a strategy of reconfiguration and progressive hollowing out of the building's interior spaces in order to allocate informal and free spaces to traditional training spaces (Schugurensky, 2007). This strategy makes it possible to create hybrid educational spaces where time spent in and out of the classroom helps to reinforce learning.

Architecture and Data

Organizing space and time has long been the privilege of the architect. The architectural project is as much an organization of spaces, uses or styles, as a projection into the future time of a place. Thinking about the future from the traces of the past puts the architect in a situation of an archivist of the possible.

But the architectural discipline has been used as a model on several occasions to develop new methods of organizing or structuring knowledge, as the following two examples illustrate: methods of memorization or computer programming or coding languages.

For example, since ancient times, the experience of architectural places has been a powerful support for our memory. As recounted by Francis Yates in *The Art of Memory*, in ancient Greece, memorization techniques called "memory palaces" were developed to train the memory of men of science. In this book, Yates relates that physical space is the basis for a mental or fictional space used as a mode of memorizing data. This "method of places" anticipates, in a certain way, the emergence of hypertext (Bush, 1945), a new mode of reading and sharing knowledge through the diffusion of non-linear texts organized by semantic or thematic affinities.

Another example of the strong relationship between architecture and organizational models can be found in the book "A Pattern Language." In this book, the English architect Christopher Alexander, in 1977, analyzes the organization of cities and their components: urban spaces. The city, because of its large scale, can be broken down into micro-public spaces of smaller scale and of elementary nature that could constitute, from a limited number of configurations, the matrix of all known cities. According to Alexander, a language of 253 "prime patterns" would resurface in all known cities. This thought, surprisingly, will later have more impact in the field of computer science, especially in the creation of object languages, than in architecture. By identifying similarities (patterns) in the complex organization of cities and by demonstrating their recurrence, this research had a considerable impact on the emergence of computer programming languages and the organization of digital data.

These protocols have gradually allowed the implementation of a common syntax for globalization—the linking of world economies—since the 1990s. They have also participated in the deployment of automatic data aggregation and querying procedures and then, in a more recent period, in the emergence of new modes of collaboration (e.g., continuous design for the industry) and social exchange (digital social networks) of a scale and scope never before envisaged. This later led to the development of global multi-service digital platforms, bringing together social networks, banking services, mobility services, and collaborative work platforms.

Digital technologies have thus infiltrated our social and cultural activities and have profoundly changed the practice of our territorial space and our habitat. Our physical and virtual activities have generated a multitude of digital traces that computer systems have consolidated into data lakes. This mass of data (Big Data) associated with the developments of machine learning (deep learning) has allowed the development of algorithms, which, by iterative improvement, are able to identify elements as complex as the level of dryness of agricultural fields in territorial satellite photos, tumor embryos in medical imagery, or to predict our desires from our online consumption practices. The larger the dataset, the more reliable the prediction accuracy, as shown by the learning curves of the calibration evaluation of these systems. In this example, by multiplying the data field by 3 (from 1000 to 3000 examples), this system increases the reliability of the results by 20% (from 70 to 90% performance) and exceeds human classification performance. This mathematical approach, awkwardly called artificial intelligence (AI), has continued the industrial revolution with the automation of an increasing number of our activities, this time not physical (industrial revolution), but especially intellectual (expert systems), with levels of precision and prediction, which we had not anticipated.

The conceptual impact of these technologies on architecture remains largely underestimated today, despite research works that have invested the field of architectural or urban design, such as the capacity of machines to think (Turing, 1951), autonomous architectural design—"Architecture-By-Yourself" (Negroponte & Friedman, 1976), neural network programming for learning (LeCun, 1991), mass customization in a digital manufacturing process (Cache, 1997), or researching the limits of AI technology creativity and empathy (Kai-Fu Lee, 2018), and automatic project design through natural language (Dall.E, Open AI, 2022).

In the reflection and production phases of the architectural project, these technologies help, replace, and displace the role of architects in several stages of the project. These evolutions participate in the transformation of the missions of architects and their work to an extent that remains difficult to quantify but whose impact is certain and important in the short term. In reaction to the emergence of these technologies, some architects prefer to neglect this evolution and lock themselves into an artistic type of architectural practice (Corbusean, neo-modern, or poetic) in resistance to this wave rather than thinking of the role of the architect as a guarantor of the quality of the building by collaborating with these systems.

Others are looking for ways to limit and restrict competition with these systems in the complexity and inconsistencies of the regulations that monopolize the profession. But can regulations stay ahead of the rapid evolution of these technologies for long? My investigation on this subject shows rather that it is in a meta-collaboration man/machine that the architects can progress and increase their creativity and productivity and facilitate the difficult coordination of the multiple actors of the architectural design and the act of building.

Architects would thus be at the center of an ecosystem of data on the design and management of buildings and cities, the importance of which is underestimated and on which the skills of architects are very partial, if not non-existent. The complexity of building construction, within teams, sometimes very large, of design, construction, and management of the life of the building reinforces the importance of the field of rich and complex data that remains to be structured and consolidated (BOS, 2021). In the 2000s, industries such as the automotive or aeronautics industry, thanks to powerful sector operators (Dassault Systems, Stellantis, Airbus, etc.), achieved complete integration of the design process in a continuous and shared mode, improving product quality and lowering manufacturing times and costs.

Today, the 3D digital model of the architectural project, often wrongly named BIM (Building information model), is rather a simple model of collaboration, even if the culture of the engineer (rationalize and optimize) and rarely in phase with that of the architect (the artistic coherence of the building-work). It allows us to articulate the design of the project through intuitive modeling, its architectural production by the intelligent plan book, and to consolidate the data field of the project during the construction. The digital data thus generated by the construction process or by a connected building are a new source of data for an epistemology of architecture.

Towards Augmented Architectural Design

What is the relationship between architectural design and digital data? It seems to us that two elements allow us to explore this question.

As mentioned above, the production of data linked to our uses since the beginning of the 2000s, by social or commercial platforms, by mobile digital devices that follow us everywhere, but also more generally by the Internet of Things, constitute huge lakes of data (called Big Data) that we have archived and made accessible. These sources, with their varied structures, constitute an excellent deposit of examples, behaviors, actions/reactions that we can interrogate to predict behaviors or answer a certain number of queries.

The second aspect is the recent maturity of deep learning, which, associated with several neural network programming techniques, allows an increasingly precise response to prediction objectives, analysis requests, or generative image or 3D models. We can cite here two examples:

–Recently, online applications of Text2image have reached efficiency and quality, as demonstrated by the series of applications such as MidJourney, Dall. E, and Stable Diffusion. Based on a semantic description, these tools imagine, draw, or generate an image based on the assembly of visual references corresponding to the query in an almost immediate way.

–ChatGPT, a new conversational agent, also achieves impressive results in terms of the reliability, quality, and completeness of the answers it generates on subjects as vast as philosophy, technology, ethics, and mathematics. As Yann Lecun mentioned in 2022, these systems are progressing but remain devoid of a common consciousness, allowing them to arbitrate or guide choices or actions.

The impact of this technology on architectural thinking is multiple: the deployment of DBD (design by data), the automatic ideation from client briefs, the generation of plans or 3D models from instructions, the optimization of systems with multiple parameters such as the cost of construction, or the optimization of energy management or ventilation.

However, the data economy that we have entered does not have a zero environmental impact. We can mention, among others, the important energy consumed to refresh the processors of the gigantic data centers that store our data, as well as our software. However, energy-efficient processors are rapidly improving, and the emergence of quantum computing especially allows us to foresee a multiplication of computing power while leading to an inversely proportional drop in the energy consumed.

Data and in particular the tracking and monitoring of environmental data is the key factor to detect changes in our environment such as: The measurement of temperatures, the measurement of winds on the surface of the earth, the measurements of the chemical composition of our atmosphere, the level of ice and snow on the earth, the retreat of forest areas and the increase of urban areas and human traffic.

These data revealed the rapid disappearance of the number of species of biodiversity, and allow us to understand the evolution of these phenomena which for thousands of years were not visible on the scale of a human generation.

Textile in Architecture and the Use of Ornament versus Structure

Draped inspiration—Sculpture of Athena, Mid-2nd century BC. Saint-Raymond Museum ofToulouse, France

Could we not say that every invented form manifests itself as the fold—crumpled, formless, and torn up, even—of an earlier form? It is, at any rate, a form of drapery."
Georges Didi-Huberman

1. In his efforts to rehabilitate Aby Warburg's concept of survival (Nachleben), it is surely no coincidence that Georges Didi-Huberman chooses the fall of drapery to project it into the phantom model of art history[1]. Survival always implies a movement of peripheralization, which comes from both an evaporation and a deposition, a fading away and an archiving. An active memory that disrupts the continuity of the history of forms. The structure of exclusion here is the movement of the cloth/veil/shroud, which offers to the gaze the nudity of the Nymph's body and, at the same time, gives the drape its formal autonomy. At the bottom of the painting, on its edge and almost outside it, the drape/fold cultivates its spectrality and residual power. Derrida writes:

1 Georges Didi-Huberman, *Ninfa Moderna,* Essai sur le drapé tombé, Paris, Gallimard, 2002.

"*Always a form against a ground, but the parergon is a form whose traditional determination is not to detach itself but to disappear, to sink, to fade, to melt precisely when it is deploying its greatest energy.*"[2]

2 Jacques Derrida, *La vérité en peinture,* Paris, Flammarion, 1978, p. 72.

This sums up the paradox of the parergon, its uncertain location: what is intrinsic to the work is somehow organically linked to what is extrinsic, in what traces its limit by bending it back and dispersing towards the inside. Like Aby Warburg before him, Didi-Huberman invites us to consider the permanence of forms and their survival, not in their capacity to fundamentally and essentially model the basis of a culture, but on the contrary, in their subtle outspreading in the background, under cover of derision and tenuousness.[3]

3 (...) what survives in a culture is the most repressed, the most obscure, the most distant and the most tenacious, of that culture. The most dead in a sense, because the most buried and ghostly; the most alive just as much because the most moving, the closest, the most impulsive." Georges Didi-Huberman, *L'Image survivante, Histoire de l'Art et temps des fantômes* according to Aby Warburg, Paris, Editions de Minuit, 2002, p. 154.

Starting with this liberated garment, this fabric that no longer clings to a body or veils anything, what emerges is an isolated, unsettled figure, whose movement no discursive or figurative project can fix, an optical blurring of the frame whose appearances/disappearances verge on the hallucinatory. The autonomy and disappearance of the figure mark the end of circumscription, as subtly underscored by the falling material—an exteriorization that distances body and material in a reciprocal movement of expulsion. In the transformation of the garment, from the clinging to the draped or from the crumpled body to the uninhabited cloth, we can observe the circulation of an intensity in which the pictorial space adopts what marked the fallen limit of the body, its clothing, as a limit of/to Representation. At the same time, as it sends or projects the drape towards the margins, confronting it with its own uselessness or emptiness, it frees its formal power. Its exclusiveness is both its exclusion and the guarantee of its persistence.

The excess (for us, ghost track) and the outside of the work (parergon) open up the formal possibility of another way that already guarantees the conditions of its dispersion and survival. Decoration, or that which stands (or falls, or is laid down) as such, is here, in more ways than one, a coincidence, and its concept a hallucination.

2. Gottfried Semper's work's current popularity prompts us to set stereotomy against tectonics, the motif against the model, as if, at its edge, the practice was revitalized by formal principles that are more essential (its essence?) and more fluid (its anthropological fluid?).

As we know, the principle of the textile essence of architecture, as stated by Semper in *Prinzip der Bekleidung*, posits the shell and its textural as primordial, preceding the structure. Architecture begins with and in ornament, or rather, in braided, twisted canvas, not with its capacity to provide shelter or put up an enclosure but with its actual fibrous quality. Mark Wigley writes as follows of Semper's thesis:

"*The interior is not defined by a continuous enclosure of walls but by the folds, twists, and turns in an often discontinuous ornamental surface.*"[4]

The discontinuity of the twisted surface comes in at several different levels here. First of all, it implies a definition of architecture as more or less indifferent to the idea of spatial circumscription. Moreover, as we will see later on, it evokes a discontinuous, riven idea of the project.

4 Mark Wigley, White Walls, *Designer Dresses, The Fashioning of Modern Architecture*, Cambridge, MIT Press, 1995, p. 11.

In referring to Semper here, we are not trying to point to some original form or formula, but rather, an ur-organization of structure and surface that would provide an ideal model for the distribution of roles. On the contrary, turning the structure/surface dialectic towards the primacy of ornament all too often imprisons this in the role of screen, mask, or intersecor.[5] What is more, it freezes the free play of its concept.

5 The veil that Alberti advised the painter to place between him and the scene to represent in order to see its outline better.

The decorative destiny of the twisted surface begins the mourning process for the superposition of motif and structure of Werkform and Kunstform, for it is the motif that structurally holds the drape. Wigley again: "Strictly speaking, it is only the decoration that is structural."[6] What matters in the braided surface is the coincidence of graphic and structural, the textile pattern that determines its internal tension and physical behavior. The fold adds a second pattern over that, a more ample and distinct one, but dependent on the primary structure. From this differential process of the surface's constitution, we can, we believe, draw out structural and formal issues that go well beyond the Werkform/ Kunstform debate mentioned above.

6 Mark Wigley, Ibid, p. 11.

3. Manuel De Landa[7] notes that it is easier to create a digital model of verbal knowledge (like "teaching" a machine to play chess, for example) than to model a savoir-faire (doing), a construction technique, a technique of manual elaboration that implies irregularity, hesitation, trial and error, internal logics of compensation, heterogeneous and programmatic condensations, assemblages unraveled and raveled up in an attempt to keep the surface as a whole together. The technique of braiding evokes this kind of complexity, these strategies of finalization. It points to the power and product of transformative intelligence, the mental and material process shared by any kind of tekhnê. The corpus of technical gestures linked with this knowledge, given as an ensemble of know-how, resonates with an overturning of limits. The art of braiding, knotting, twisting, and curling motifs, the oscillation of the thread around the fabric, are as much the echo of infinitely inversed/reversed figures as they are of dreams of spaces without seams, without joints, whose weave absorbs in its density the linearity of its contours.

7 Manuel De Landa, "Les Philosophies de conception : the case of the modelization software", in *Verb, architecture boogazine*, Barcelone, ACTAR, 2002, p. 134.

Braiding is the product of a general economy, a differential principle of diminishing interstices, and the illegibility of its constituent logic. Just like a carpet covered with decorative patterns using a simple

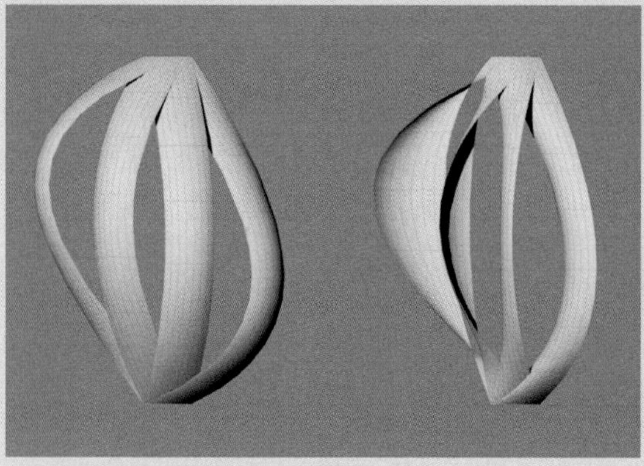

3D modeling research on woven geometry

Weaving digital and physical prototypes

movement whose repetition and recurrence proceeds by gradual complication in time and space[8], so the braided surface tends towards the loss of the figure, to the phantomization of the process of elaboration, to a general specto-graphy. An internal pulsation of the surface, both graphic and structural, blurs the legibility of its project, of that which, initially programmed, would have allowed for its braiding. The condensation of knots and the saturation of the weave create a zone of indiscernibility or undecidability whose amplitude we need to cultivate.

8 E.H. Gombrich, *The Sense of Order - a Study in the Psychology of Decorative Art*, Londres, Phaidon Press, 1979, p. 67.

In opposition to the obvious regularity of braiding stands the abruptness of technology, the transformation of the project, which, in more ways than one, is the betrayal of a memory, of a preceding utterance engaged in the mnemic movements of resurgence, recurrence, division, and anamnesis. As a savoir-faire, insofar as it is the projection of knowledge in the process of doing, the ability to braid organically integrates incident into its temporal and spatial unfolding. This is what defines its eminently event-like character. The braided surface is made by unmaking, by countering—not systematically but when opportune, in context—the programmed form of its knots.

We might wonder to what degree what is programmed is a certain kind of blindness, a reduction of the area of projection as the limit of foresight while running counter to a nodal topology. A strategy that would impose a back-and-forth movement between the eye and the hand where one would phantomize the other in a contradictory and compensatory way.

All this might encourage us to rethink the timeliness/timelessness of the "hand-made" and the phantomization of decoration in the light of its disappearance, in the age of its technological reproducibility, on the threshold of Modernity. This is not with a view to its rehabilitation, which is already well underway, but within the limits of the concept itself. The rehabilitation of decoration as decoration throws a veil over the greater activity of its forms—what Didi-Huberman calls its residual power. It ignores the parergon's knotted topology and the energy it dissipates inwards. The praise of decoration as decoration does not see that, on its surface, an organization of forces is being played out that already is twisting program/context/form into a totally new disposition.

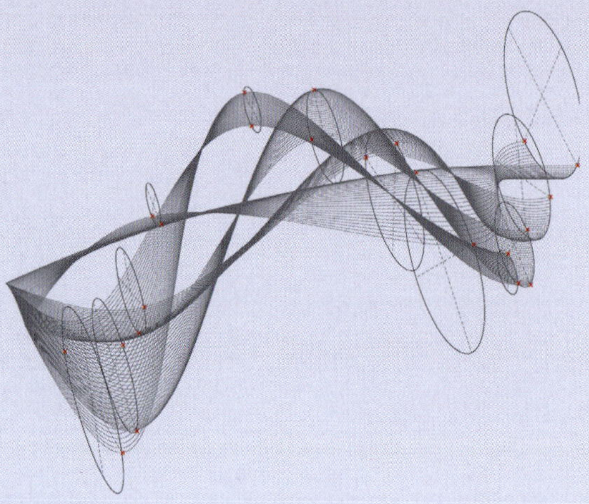

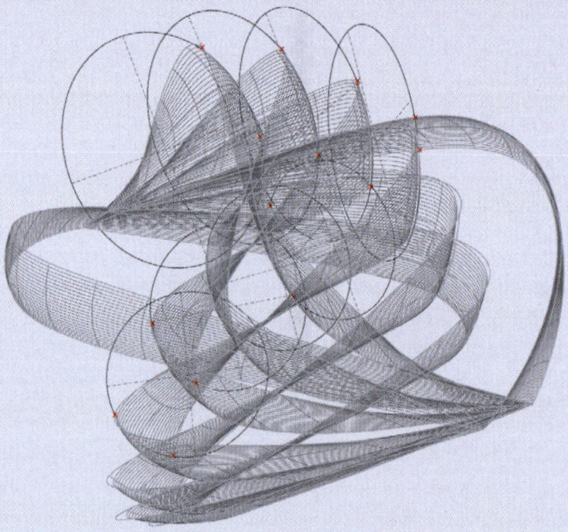

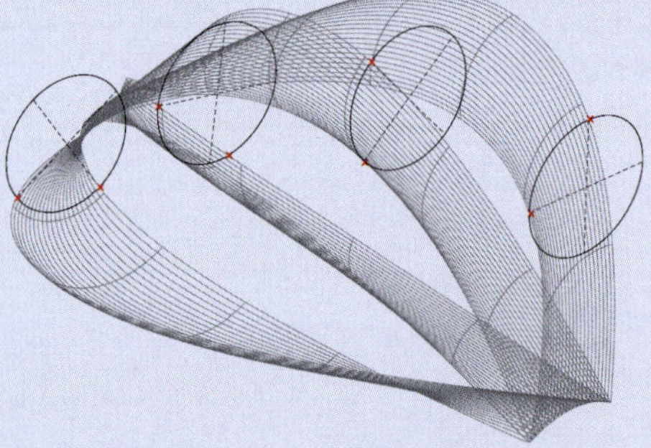

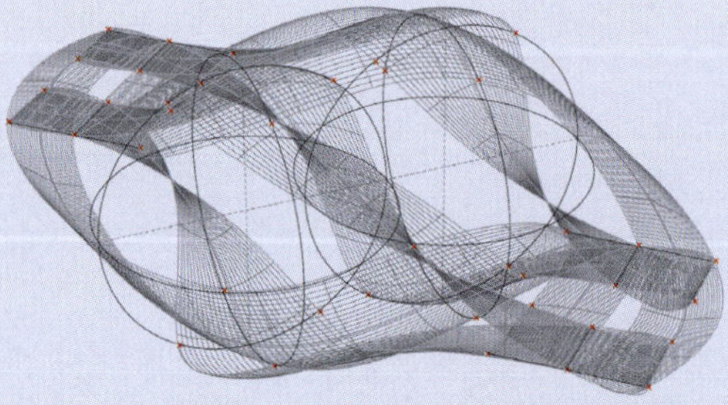

Draped inspiration—Sculpture of Athena, Mid-2nd century BC. Saint-Raymond Museum ofToulouse, France

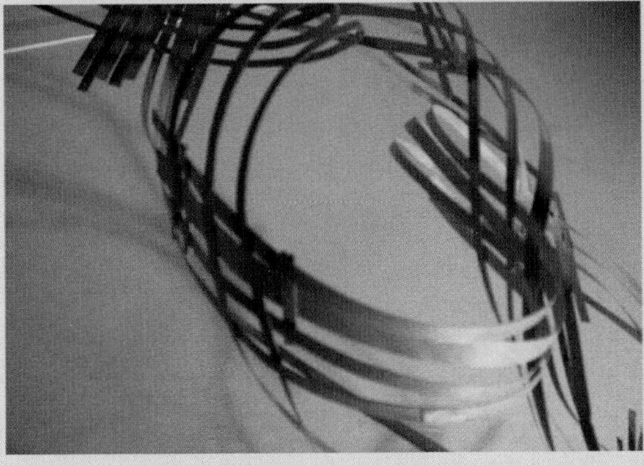

GHOST TRACK

Venice Biennale Exhibition, 2004

It is easier to create a digital model of verbal knowledge (like "teaching" a machine to play chess, for example) than to model savoir-faire or making, a technique of manual elaboration that implies irregularity, hesitation, trial and error, and internal logics of compensation in an attempt to keep the surface as a whole together.

The technique of braiding evokes this complexity, where the hands interact with the form in each of the knots. It points to the power and product of transformative intelligence, which is a mental and material process shared by any technique.

In the art of braiding, knotting, twisting, and curling motifs, the oscillation of the thread around the fabric generates an incessant inverted and reversed figure as they are dreams of spaces without seams, where the weave loses in its density of the surface the linearity of its contour.

Like a carpet covered with ornamental patterns, the woven surface uses a simple movement, in which repetition and recurrence proceed by gradual complication in time and space. Architecture begins with and in ornament, or rather, in the woven mat, not with its capacity to provide shelter or put up an enclosure but with its actual fibrous quality. The interior is not defined by a continuous enclosure of walls but by the folds, twists, and turns in an often discontinuous ornamental surface. The twisted surface's ornamental destiny begins the superposition of form and structure.

Ornament emerges within the structure.

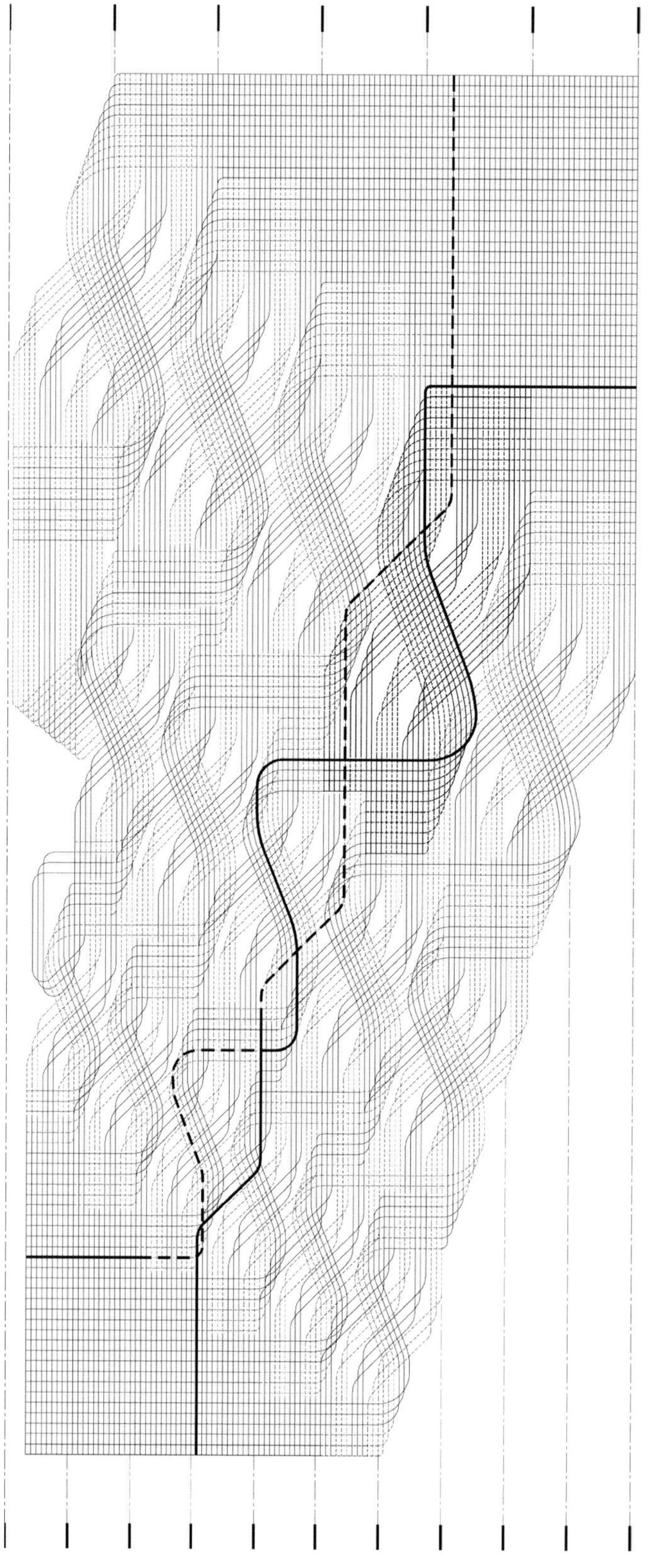

Hand-weaving instruction diagram

Right: Architectural scale woven prototype, Orleans Frac Collection 2004, dZO

LIVING CANOPY

Auditorium, Cinema and Exhibition Hall

Saint-Cyprien, 2009

Trees have always been a profound source of inspiration for me. Though incredibly complex, their structures emerge from simple yet powerful generative rules, evolving seamlessly in time and space. The efficiency of their architecture is not dictated by rigid repetition or optimization, as in modern engineering, but rather by redundancy and differentiation, principles that allow them to adapt, grow, and thrive.

This organic approach has guided the design of the new Saint-Cyprien Auditorium, located in the heart of an open park. Here, the interior and exterior merge, creating an environment that feels both relaxed and open—an architecture that invites movement, interaction, and a sense of ease.

Inspired by the silhouettes of sycamores, acacias, oaks, and poplars that populate the site, the auditorium's form echoes the intricate branching patterns of tree canopies. Serero Architects developed a custom computational script to generate a façade composed of non-repetitive, non-standard components. Despite its seemingly irregular appearance, the roof structure is based on simple geometric rules, allowing for variation and organic fluidity between its elements.

An Architecture in Continuity with Nature

The Saint-Cyprien Auditorium is not conceived as an isolated architectural object detached from its surroundings. Instead, it establishes a continuous dialogue with the park, integrating the landscape within the building itself.

A double concrete shell creates the impression of foliage-like embroidery, pierced with egg-shaped perforations reminiscent of the dappled light filtering through tree leaves. This canopy-like outer shell functions as a protective layer, shading the lobby and auditorium from excessive sunlight. The inner shell, composed of glass and concrete, regulates thermal comfort, ensuring natural ventilation and controlled heating. Visitors enter the hall by following a concrete "pebble" nestled beneath this architectural foliage.

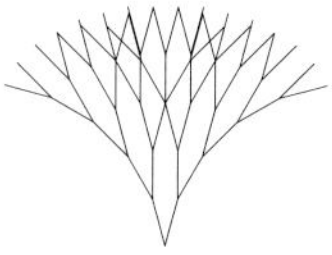

Division : 2
Length reduction : 1
Angle : 30

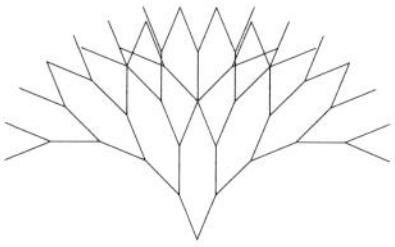

Division : 2
Length reduction : 1
Angle : 45

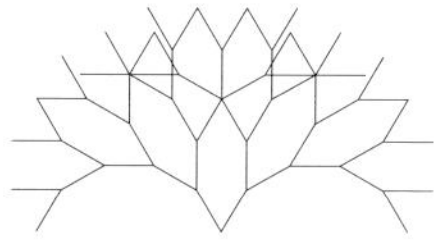

Division : 2
Length reduction : 1
Angle : 60

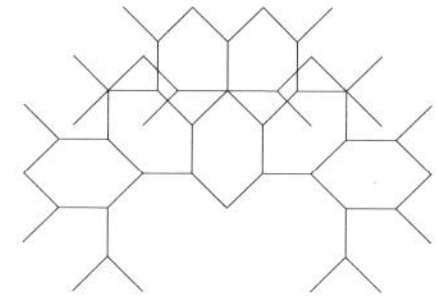

Division : 2
Length reduction : 1
Angle : 90

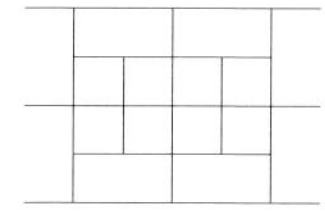

Division : 2
Length reduction : 1
Angle : 180

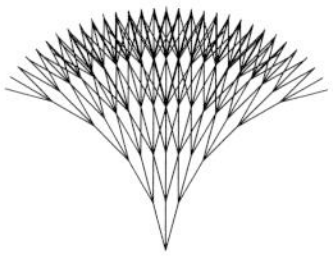

Division : 3
Length reduction : 1
Angle : 30

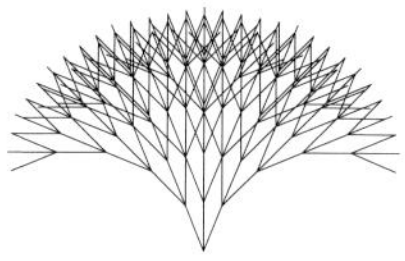

Division : 3
Length reduction : 1
Angle : 45

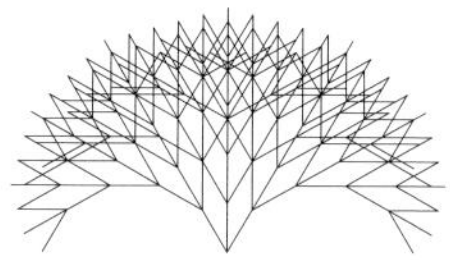

Division : 3
Length reduction : 1
Angle : 60

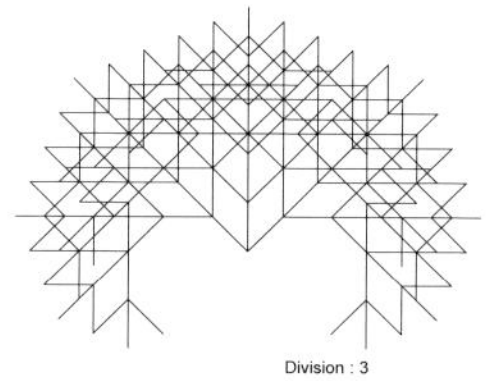

Division : 3
Length reduction : 1
Angle : 90

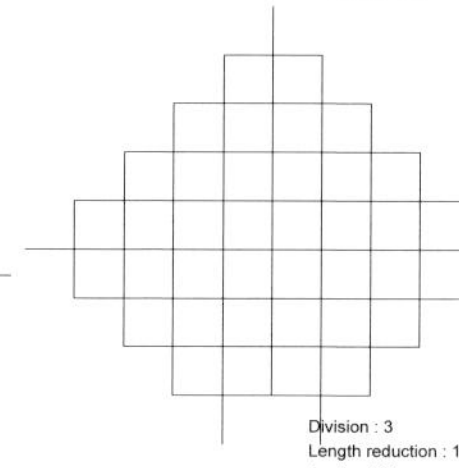

Division : 3
Length reduction : 1
Angle : 180

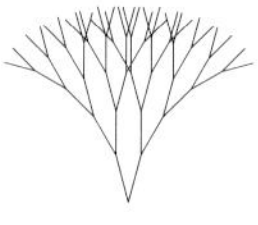

Division : 2
Length reduction : 0,9
Angle : 30

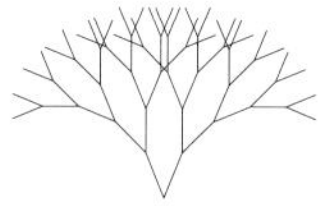

Division : 2
Length reduction : 0,9
Angle : 45

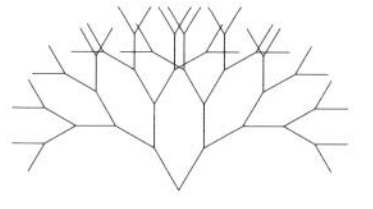

Division : 2
Length reduction : 0,9
Angle : 60

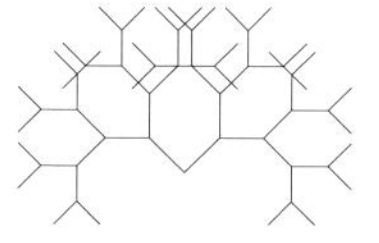

Division : 2
Length reduction : 0,9
Angle : 90

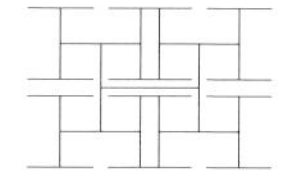

Division : 2
Length reduction : 0,9
Angle : 180

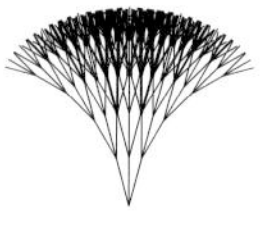

Division : 3
Length reduction : 0,9
Angle : 30

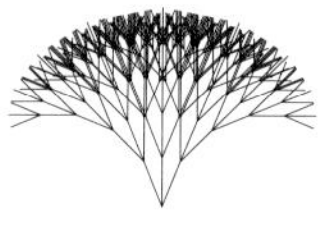

Division : 3
Length reduction : 0,9
Angle : 45

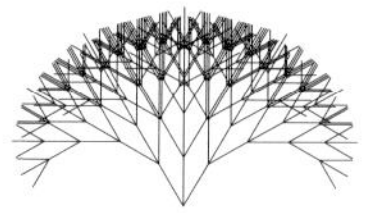

Division : 3
Length reduction : 0,9
Angle : 60

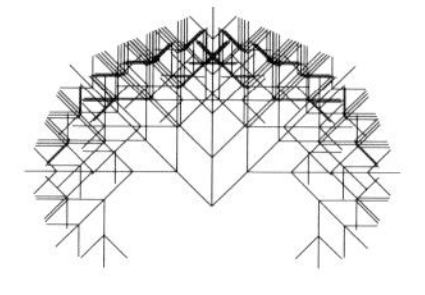

Division : 3
Length reduction : 0,9
Angle : 90

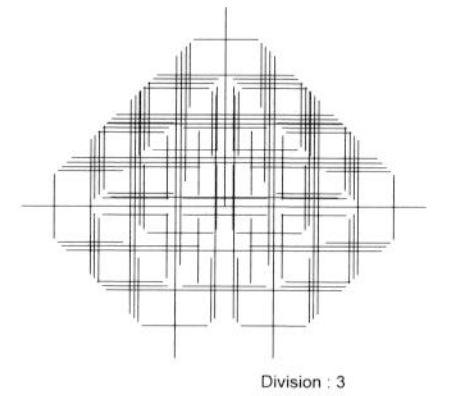

Division : 3
Length reduction : 0,9
Angle : 180

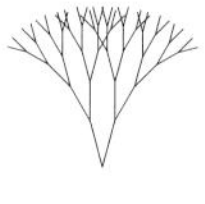

Division : 2
Length reduction : 0,8
Angle : 30

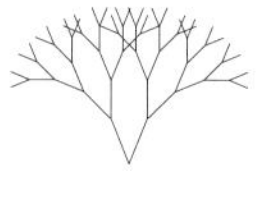

Division : 2
Length reduction : 0,8
Angle : 45

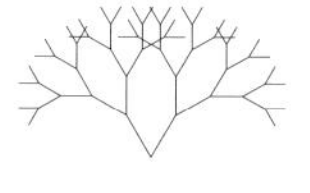

Division : 2
Length reduction : 0,8
Angle : 60

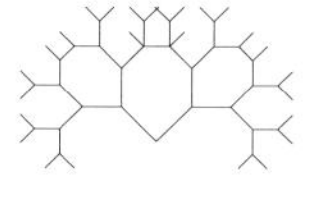

Division : 2
Length reduction : 0,8
Angle : 90

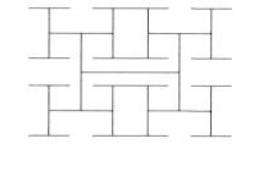

Division : 2
Diminution longuer : 0,8
Angle : 180

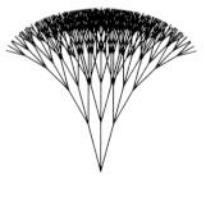

Division : 3
Length reduction : 0,8
Angle : 30

Division : 3
Length reduction : 0,8
Angle : 45

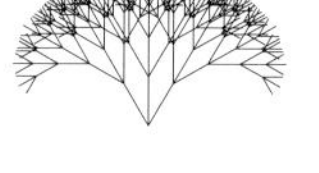

Division : 3
Length reduction : 0,8
Angle : 60

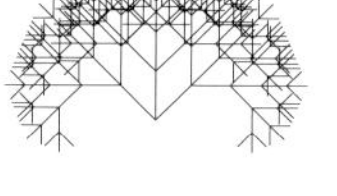

Division : 3
Length reduction : 0,8
Angle : 90

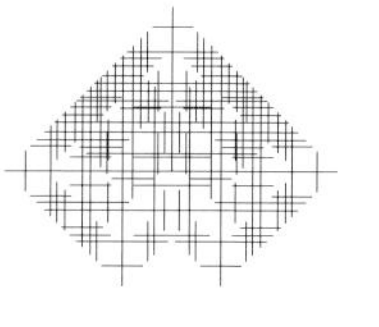

Division : 3
Diminution longuer : 0,8
Angle : 180

Branching algorithmic studies

A Responsive, High-Performance Envelope

The shed-like roof acts as a living skin, actively regulating the atmosphere within the auditorium:

–Natural Lighting Adaptation: Natural light floods the interior throughout the day. As evening approaches, integrated lamps gradually take over, seamlessly transitioning from daylight to artificial illumination.
–Passive Ventilation & Thermal Comfort: The ventilation system harnesses passive cooling in summer while retaining heat in winter, minimizing energy consumption.
–Renewable Energy Integration: Photovoltaic panels on the southern-facing roof generate part of the building's electricity, ensuring a sustainable energy contribution.

Beyond its functional performance, this embroidered canopy becomes a distinctive architectural signature, defining the auditorium's identity. The interplay of concrete oculi and translucent honeycomb panels creates a dynamic, ever-changing visual effect, reinforcing the idea of architecture as a living, breathing organism.

An Auditorium Designed for Acoustic Excellence

The interior of the auditorium has been meticulously crafted to achieve optimal acoustics, ensuring an immersive auditory experience for diverse performances:

–Wooden Slat Walls with Acoustic Treatments: The interior walls, clad in wooden slats and specialized acoustic materials, are carefully tuned to accommodate both live performances and cinematic screenings and deliver exceptional sound quality.
–Geometrically Complex Ceiling for Sound Diffusion: The intricate ceiling design enhances sound diffusion, ensuring that audio is evenly distributed throughout the space, creating a rich and immersive experience for every audience member.

This new Saint-Cyprien Auditorium is architecture in symbiosis with nature, borrowing from the intelligence of tree structures to create a fluid, adaptive, and sensorially rich space. More than a performance venue, it is an organic extension of the park, where light, air, sound, and structure harmonize to redefine the relationship between architecture and its environment.

Main hall computer light studies on June 21st and December 21st

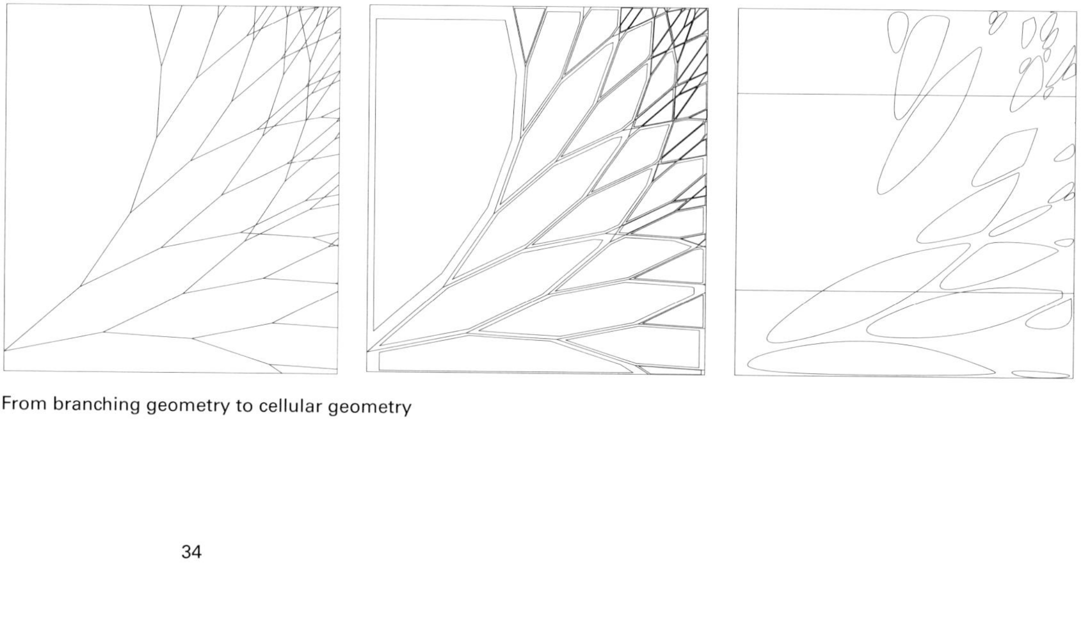

From branching geometry to cellular geometry

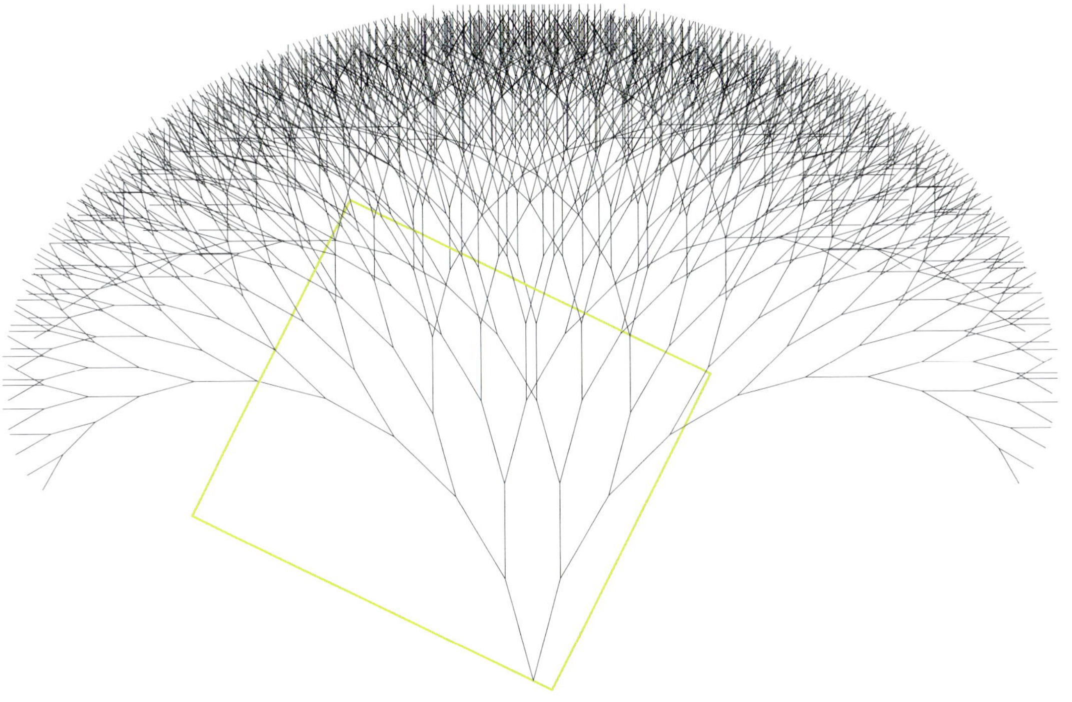

Selected branching geometry for the project

Roof structure patterns study model

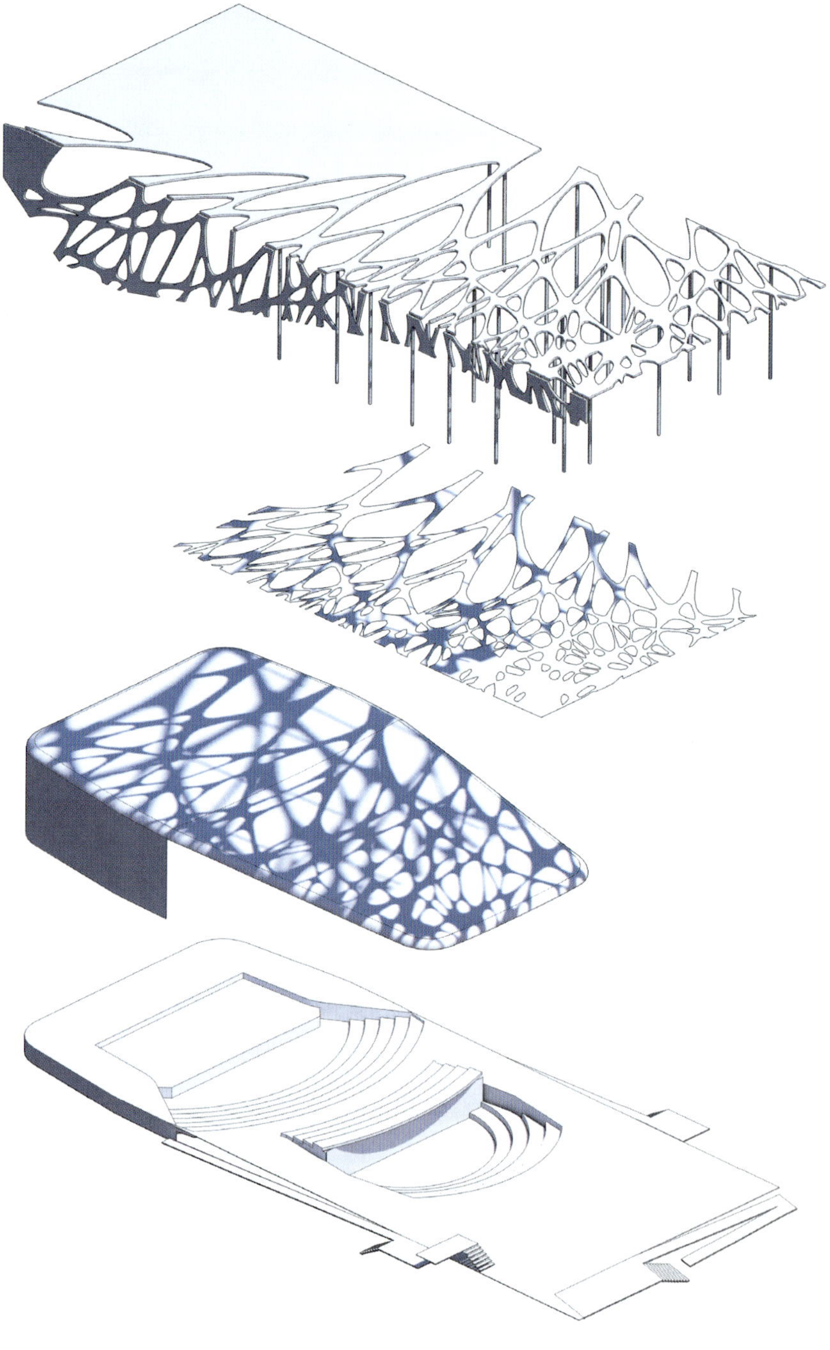

Axonometric view concert hall, openwork ceiling, and concrete shell

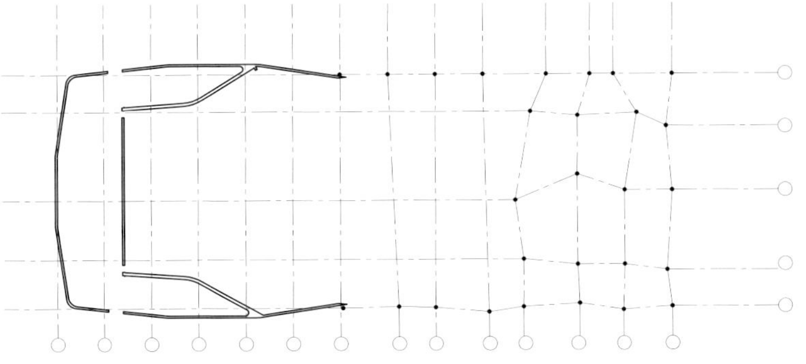

Structural plan integrating existing site trees

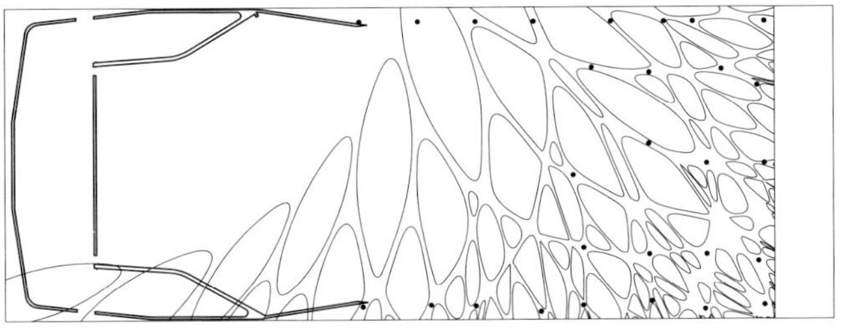

Concrete outside shell

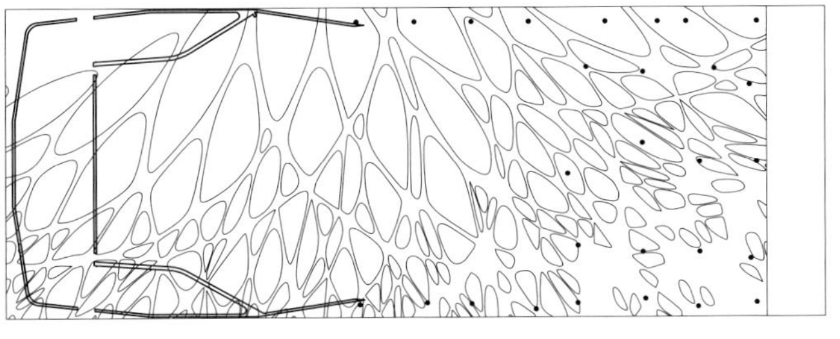

Openwork plaster ceiling

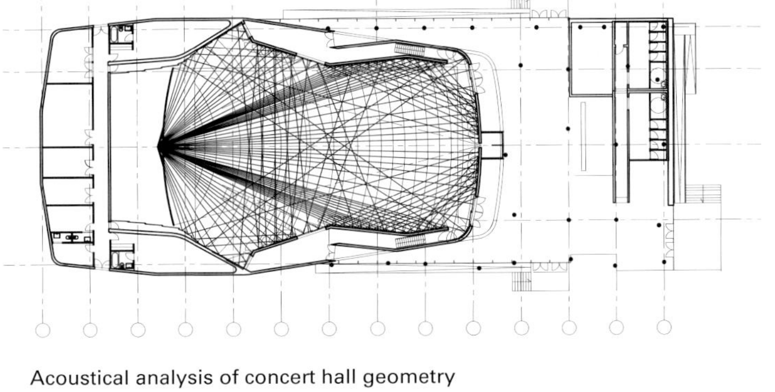

Acoustical analysis of concert hall geometry

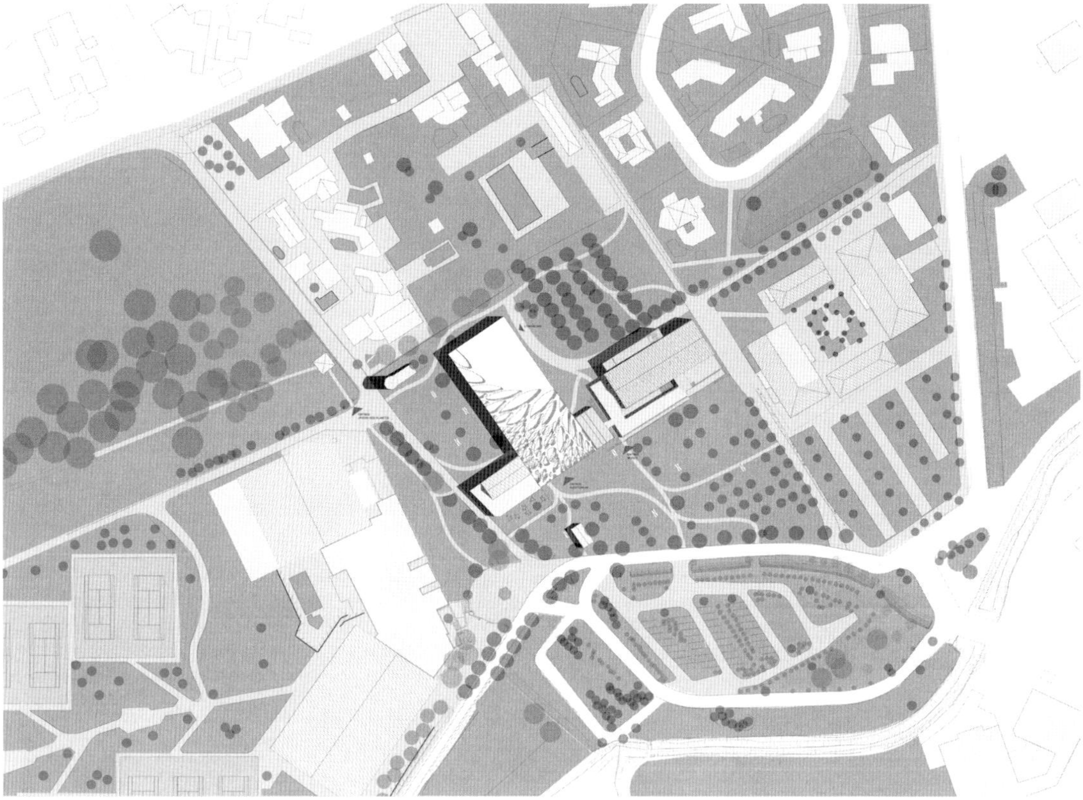

Site plan

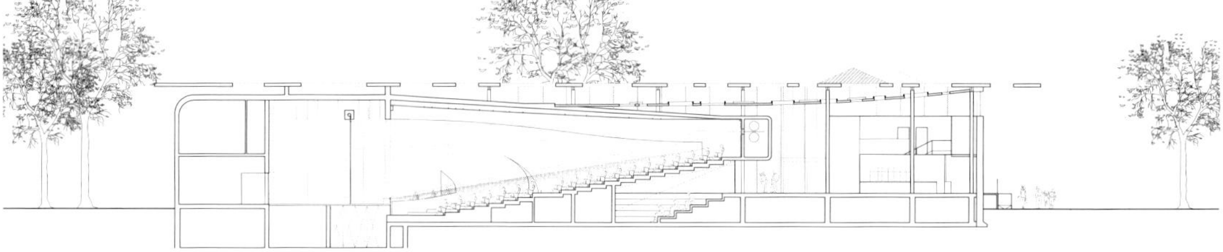

Building sections

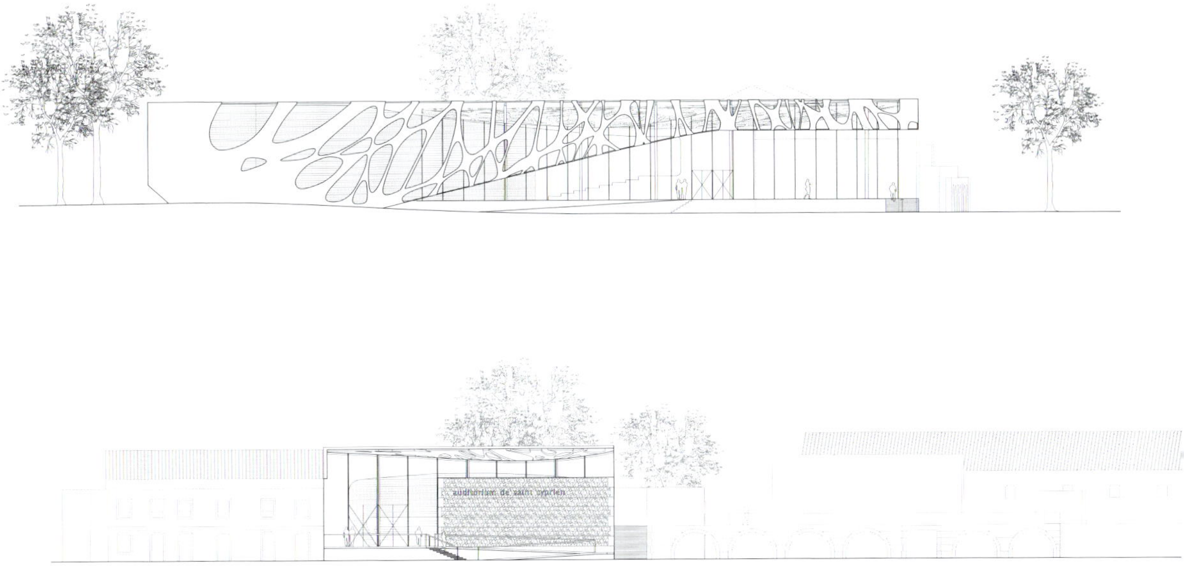

Building elevations

Concert hall interior view

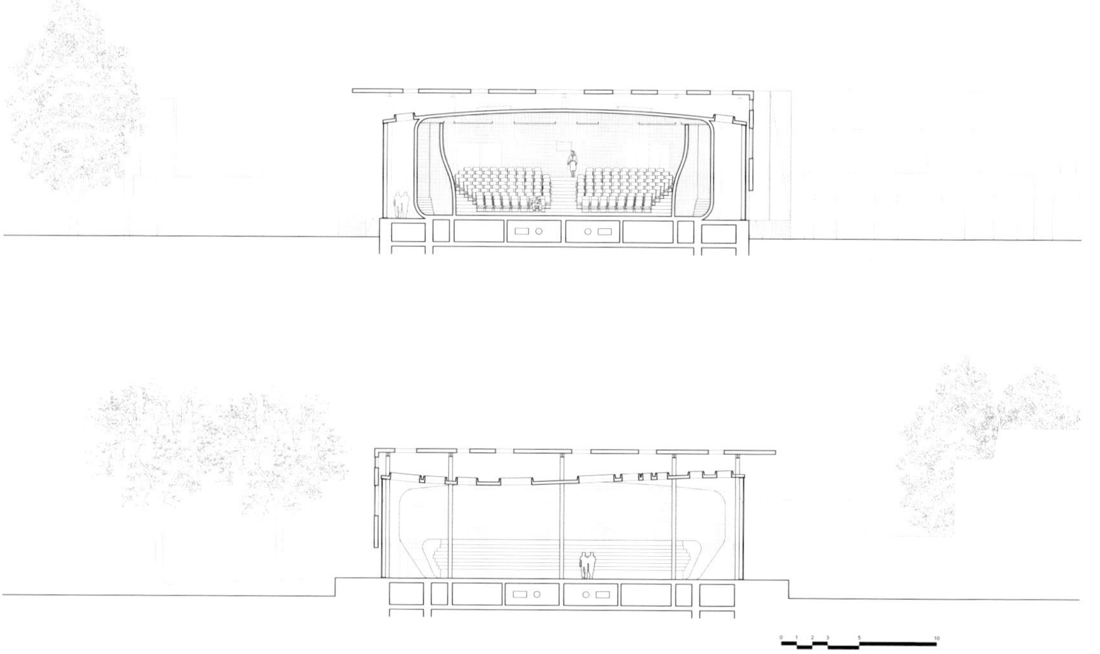

Section of concert hall

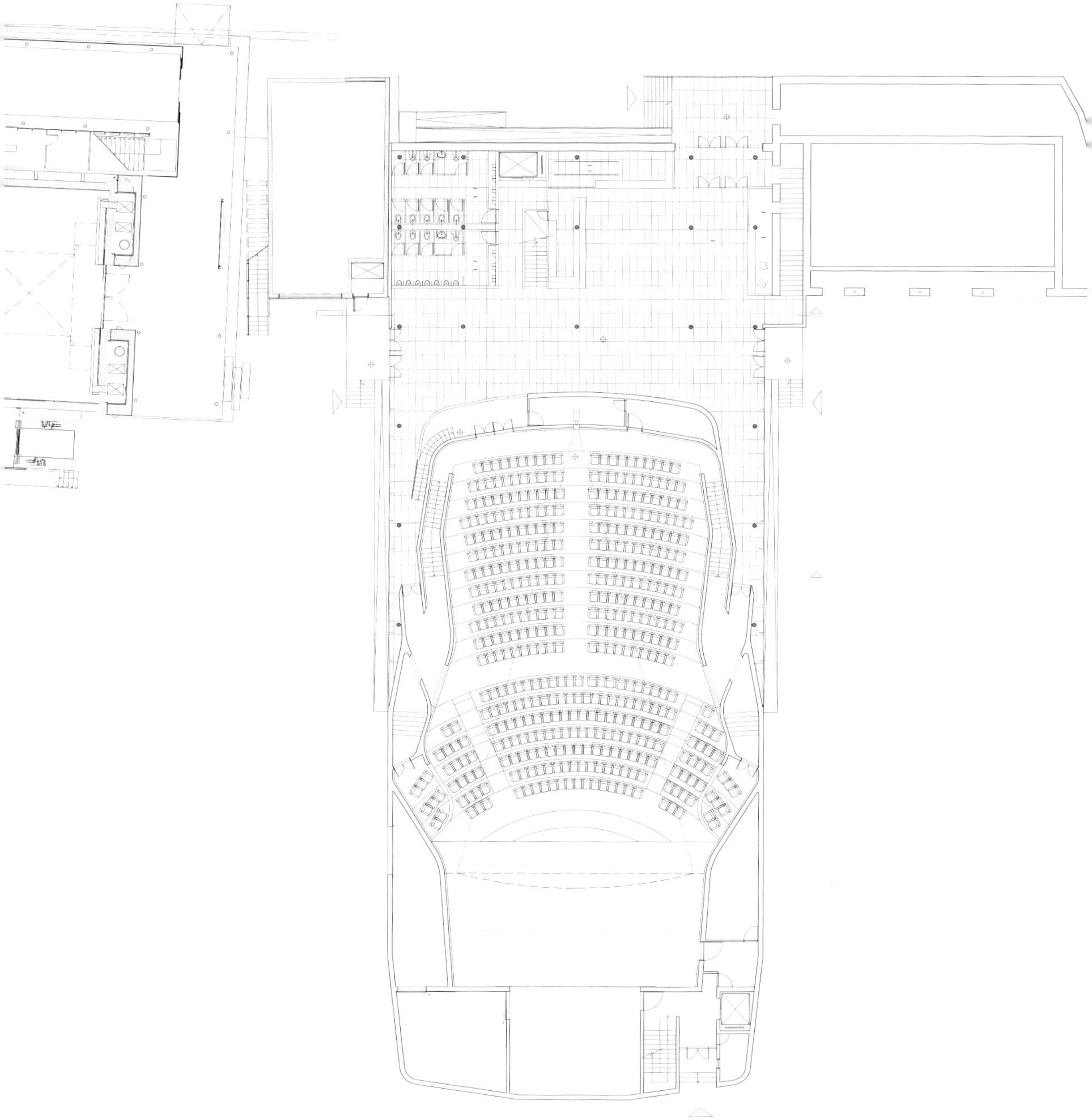

Ground floor plan

The Experience of Sound

Villa Medici, Rome

I was in the dark. A deep, enveloping black night—both unsettling and unfamiliar. A darkness unlike any I had ever known. Though the night was slowly receding, its dense obscurity remained intact, unyielding. I had no light to guide me. No streetlamp on the horizon, no candle, no glowing window. Leaving my studio at the Villa Medici, after a long night of work, I moved forward cautiously, letting my eyes adjust, attempting to navigate the void.

Step by step, I became aware of my own footsteps, their echoes reverberating in an invisible courtyard. I listened closely, discerning the direction of each sound, its distance, its decay. In my mind, I began reconstructing the space around me—the density of a hedge, the proximity of a wall, the vastness of a courtyard. Then, the subtle murmur of a fountain, then another, and another—an alignment of water sources forming an invisible axis. Instinctively, I connected it to the great transverse axis of the Villa Medici. Suddenly, a profound silence stretched before me, an abyssal void. And there, emerging from the darkness, I discovered the skyline of Rome, punctuated by domes, multiplying into the horizon.

Night after night, leaving my studio at the western edge of the garden, I wandered through the depths of this blackness. For the first time, I experienced space not through vision, but through sound. I learned to read the resonance of monumental walls, the echo of a travertine façade, the vibration of a brick loggia, the soft absorption of the umbrella pines. Each night, I mapped the Villa Medici through its acoustic presence. I delighted in decoding its soundscape, getting lost only to find my way again through sound alone.

This was the moment when my fascination with sound began—a revelation that architecture could be understood not just as a visual discipline but through its capacity for resonance, for sonic imprint.

From Perception to Experimentation

At that time, I was developing acoustic devices capable of reshaping the sonic environment of the grand reception hall of the Villa Medici. Rather than designing new architectural forms, I chose to explore sound itself—its movement, its spatialization, its interaction with built space. Using acoustic simulation software, I studied and analyzed the propagation of sound waves, not merely to simulate them but to decode their underlying logic—how frequencies and resonances could program the organization of materials, define new structural principles, and generate intelligent spatial configurations that respond to acoustic demands.

Traditionally, domes have been architectural solutions for covering vast spaces, their geometries shaped by the constraints of material resistance. My stay in Rome offered me the freedom to challenge this tradition, to imagine a dome that would transcend its structural role and become a dynamic acoustic device.

The Variable Configuration Acoustic Dome

In collaboration with an Italian mathematician, an acoustician, and a programmer, we developed a new way of visualizing acoustic behavior—a method that allowed us to generate acoustic fields as projective foundations for architecture itself.

Installed within the 14-meter-high reception hall of the Villa Medici, the Variable Configuration Acoustic Dome mirrors the room's vaulted ceiling. This experimental structure, suspended above the space, is designed to adjust and transform the hall's soundscape in real-time.

By altering its position and degree of folding, the dome actively reshapes the propagation and diffusion of sound. Here, space is not passive—it becomes a participant in the performance, an interactive medium. With each concert, conference, and voice that fills the hall, the geometry of the room adapts, shifting the acoustic conditions in response to the event and its audience.

The dome is composed of a tessellated surface, where an iterative system of folds generates an evolving harmony between music and architecture. Its shape is conceived as a convergence between digital acoustic computation and physical material deployment.

Rather than defining a fixed form, the project embraces variability, embedding the very principles of transformation within its geometry. It operates at the intersection of ornament and acoustics, allowing for an active exchange between architecture, music, and perception.

The Acoustic Cupola creates a fluid, dynamic relationship between the space, the work, the audience, and its immediate environment. It is not merely an architectural installation but a resonant instrument, a structure that listens as much as it shapes the sound it receives.

VARIABLE GEOMETRY ACOUSTICAL DOMES

Rome, 2006

Music, both a tangible and intangible phenomenon, serves as the foundation for this spatial intervention within the grand reception hall of the Villa Medici.

Suspended in this majestic 14-meter-high space, the Variable Configuration Acoustic Dome is an experimental device that mirrors the room's vaulted ceiling. Designed to dynamically adjust its geometry, the installation transforms the space's acoustic properties, modifying both its sound environment and its reverberation response.

By altering its position within the hall's volume and adjusting its degree of folding, the dome reshapes the way sound propagates and diffuses. Here, the space is no longer a passive container for music—it becomes an interactive medium where musical compositions are embedded within an adaptable form. With each performance, each audience, and in real-time, the room's geometry evolves to engage in a dialogue with the music.

The nature of the event—be it a chamber music concert, a conference, or a vocal performance—directly informs the spatial calculation that governs the dome's movements. The relationship between architecture and sound is continuously redefined as the Acoustic Cupola bends, unfolds, and adapts. The tessellation of its surface, combined with an iterative folding mechanism, allows for a dynamic synchronization between music and spatial form.

This project does not seek to define a fixed form but rather an adaptive system, embedding the principles of its transformation within its geometry. It exists at the intersection of ornamental form and acoustic performance, creating a dialogue between structure and sound propagation. The dome establishes a flexible, evolving relationship with the existing architecture, the musical work, and the audience—blurring the boundary between space, sound, and perception.

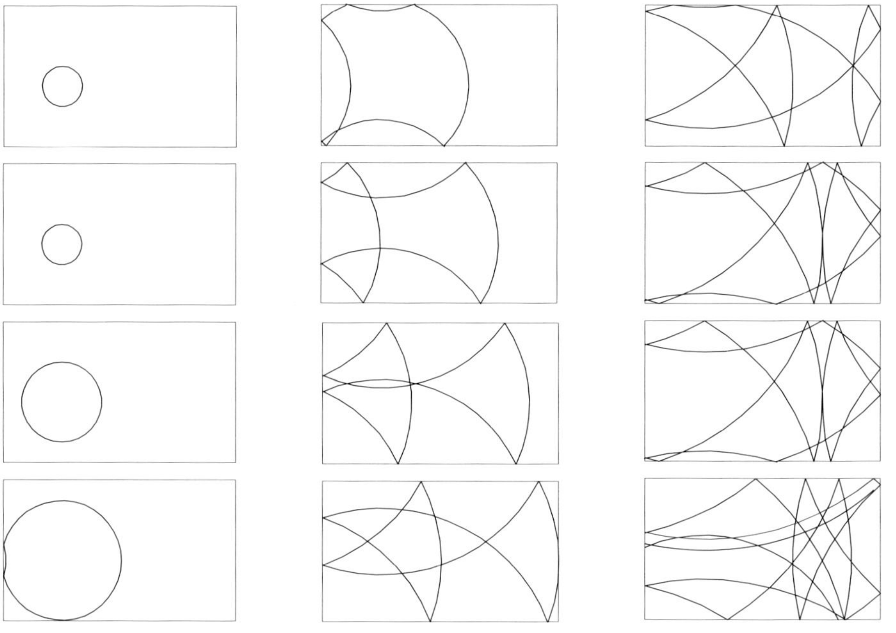

Parametric model for sound wave dissipation
in a rectangular space

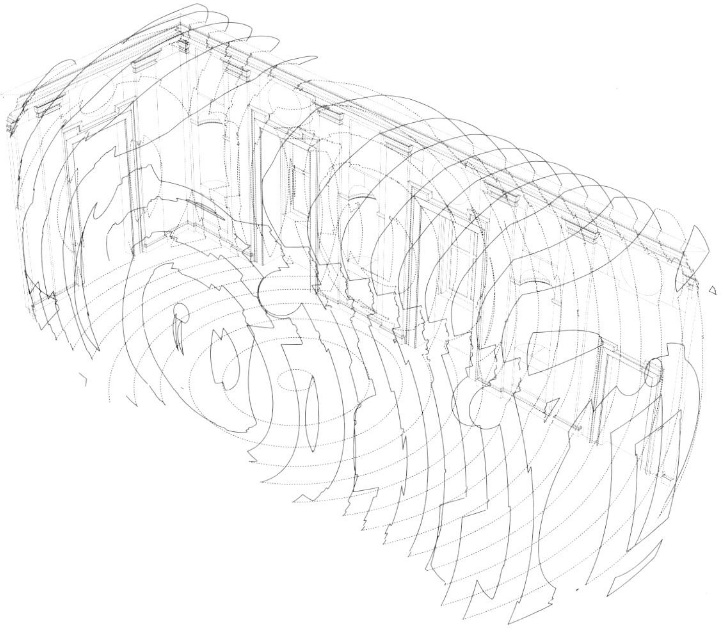

Axonometric view of sound wave intersecting
the architectural volume of Villa Medici's Great Hall

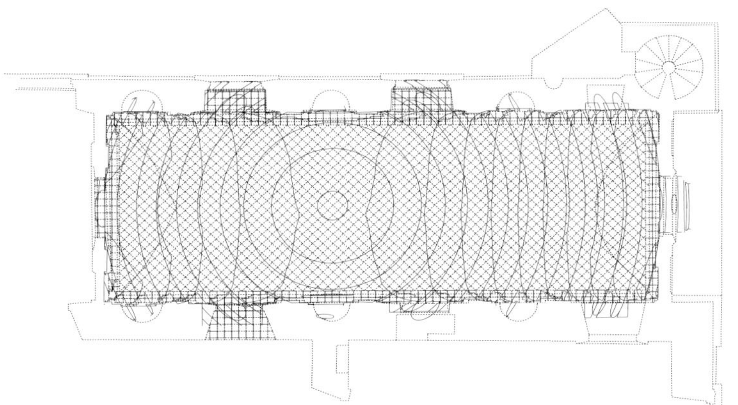

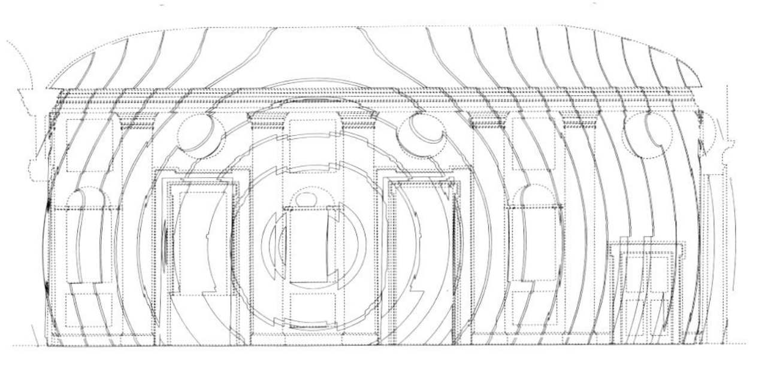

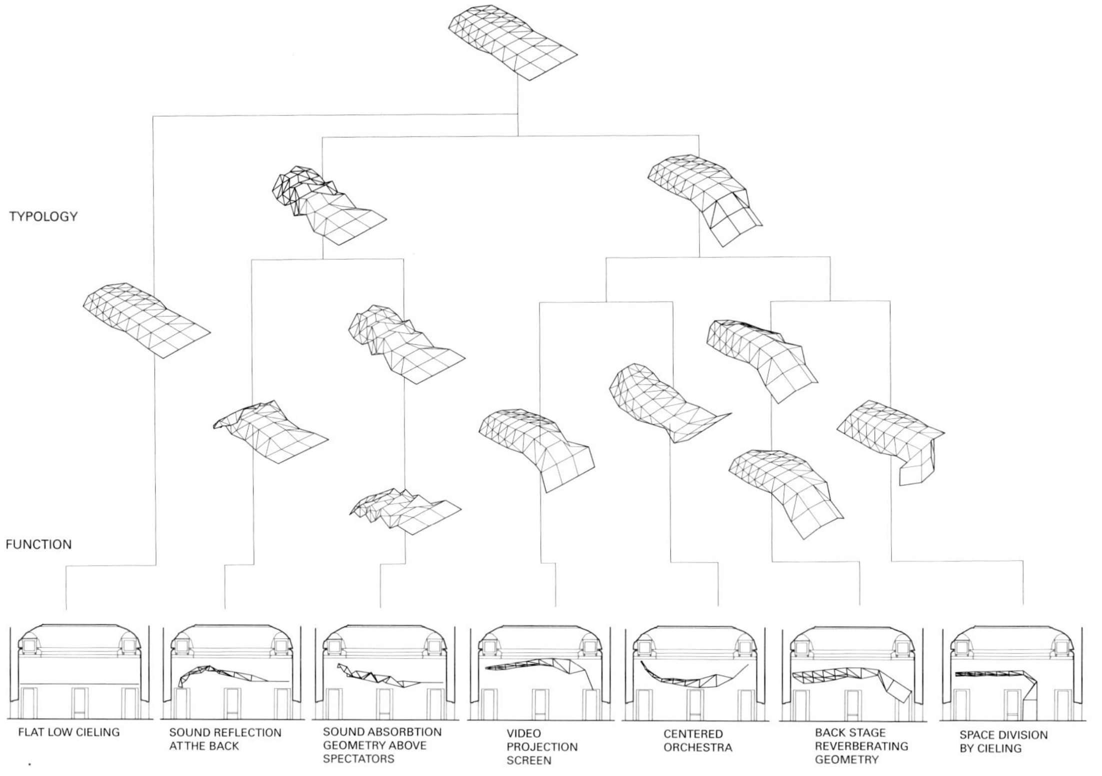
TYPOLOGY
FUNCTION
FLAT LOW CIELING
SOUND REFLECTION AT THE BACK
SOUND ABSORBTION GEOMETRY ABOVE SPECTATORS
VIDEO PROJECTION SCREEN
CENTERED ORCHESTRA
BACK STAGE REVERBERATING GEOMETRY
SPACE DIVISION BY CIELING

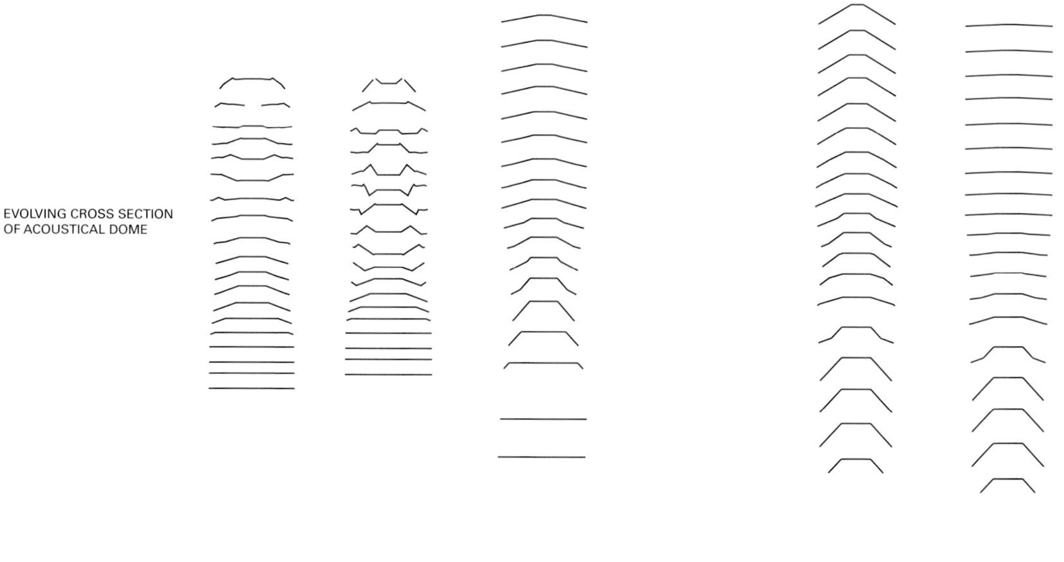
EVOLVING CROSS SECTION OF ACOUSTICAL DOME

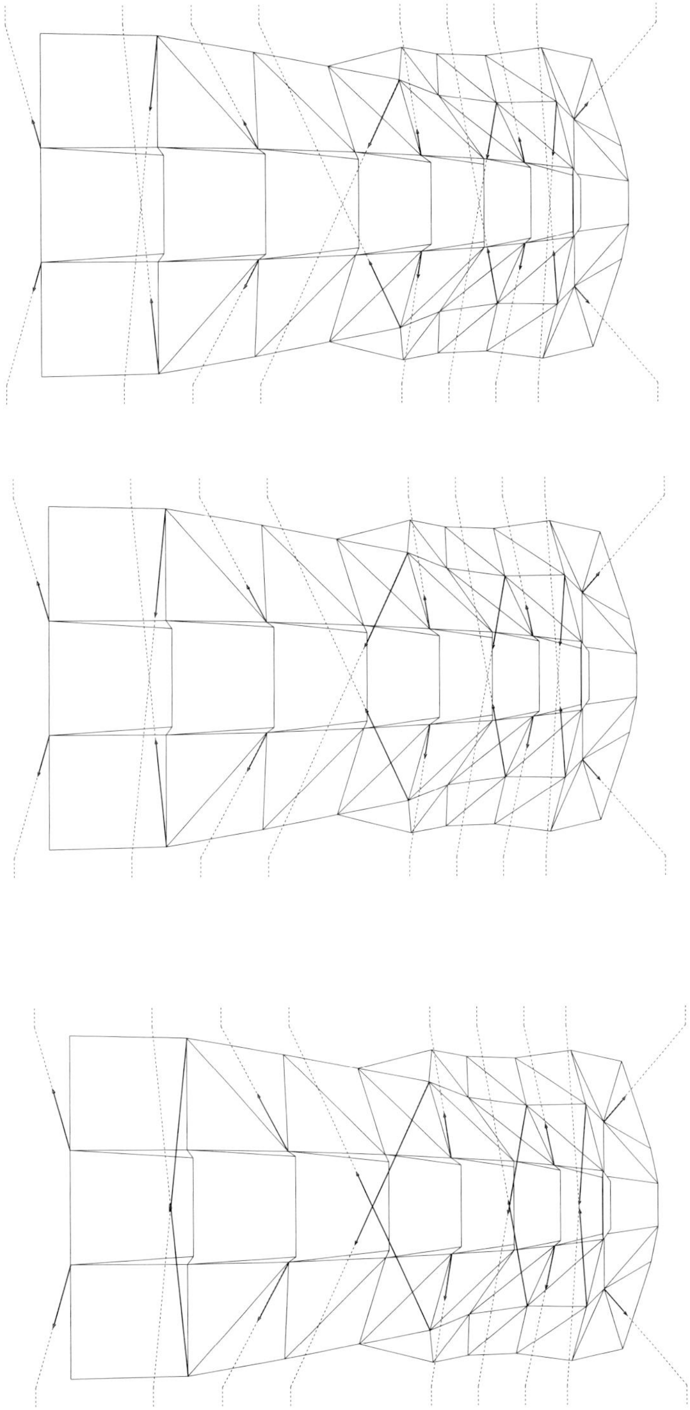

Planar geometry of variable geometry dome

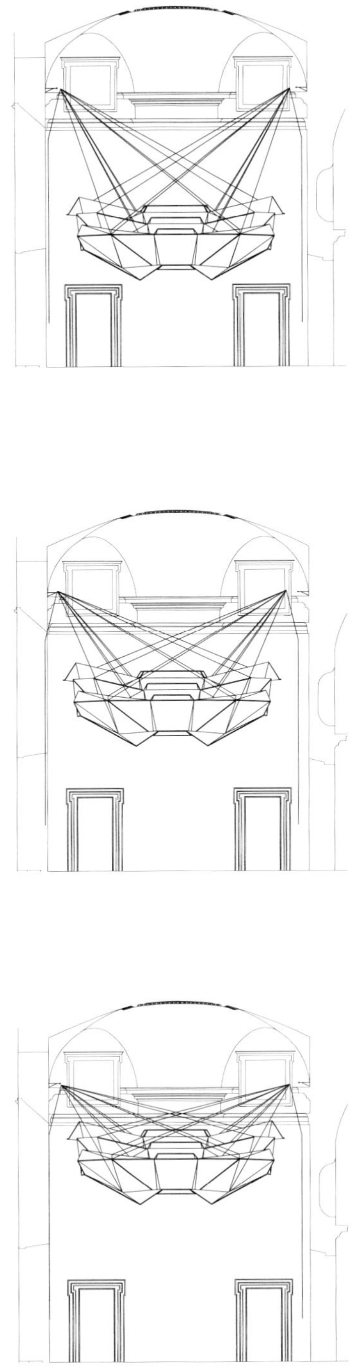

Sectional dome position

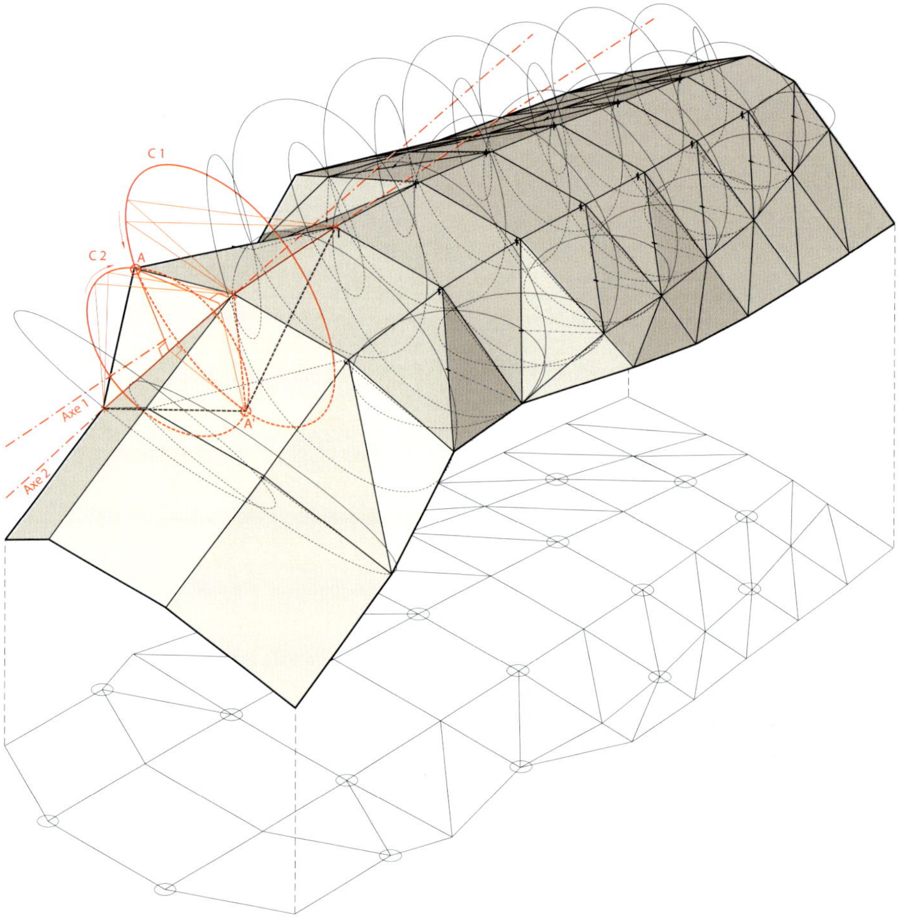

Mobility levels of acoustical dome geometry

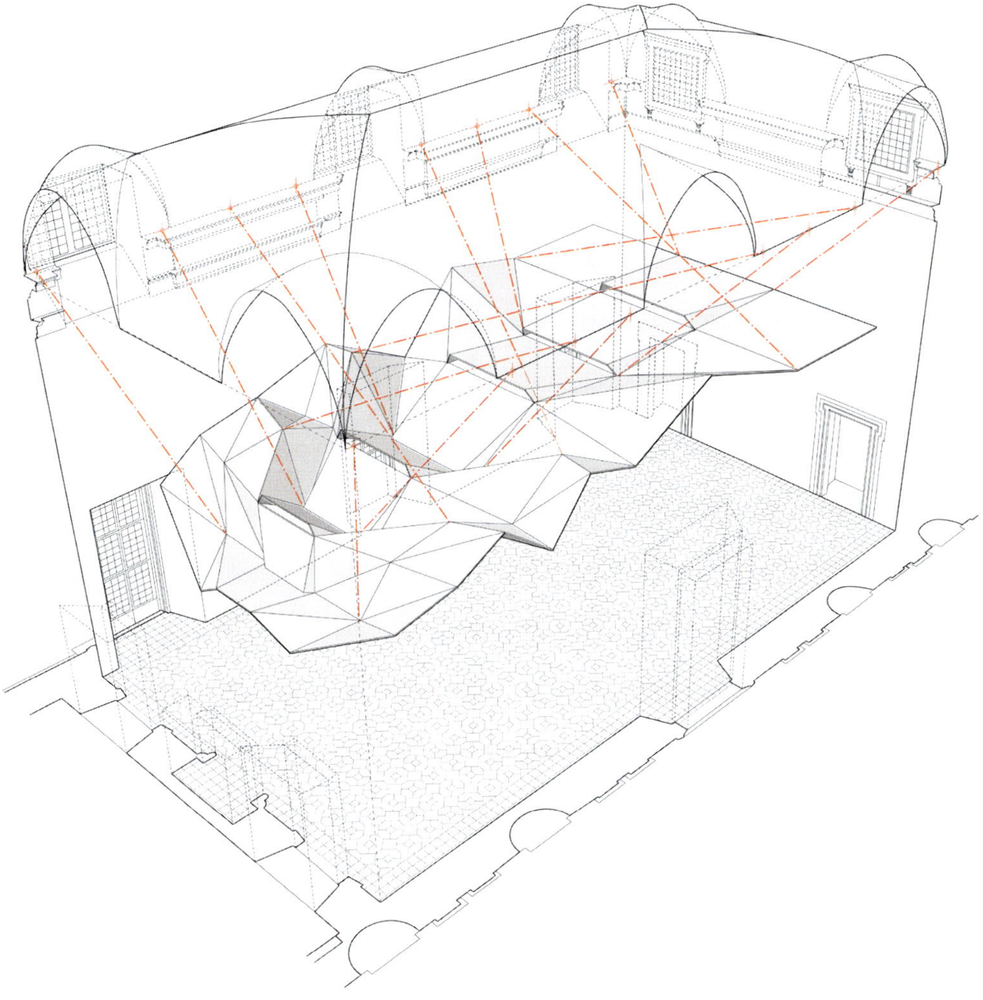

Axonometric view of Acoustical dome installed
in Grand Salon, Villa Medici

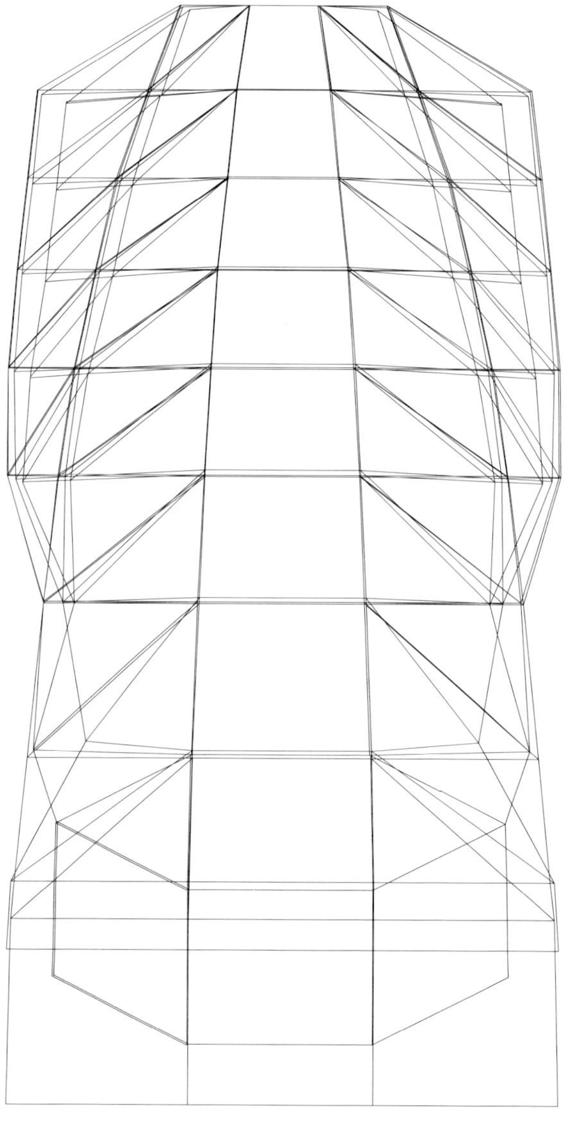

Projected geometry in plan and dome ceiling view

Planar geometry of variable geometry dome

Villa Medici—Loggia installation of variable geometry dome

ROBERT DOISNEAU CULTURAL CENTER

Meudon-la-Fôret, 2015

Tree logic Architecture

This project draws inspiration from nature, and more precisely from the tree—not merely as a symbolic reference but in its fundamental relationship with architecture. It embodies the same structural intelligence, balancing envelope and framework, redundancy and optimization, strength and fragility, singularity and diversity.

At the heart of the design is the dialogue between the forest's natural landscape and the rigorous, minimalist volumes of the existing buildings designed by architect Fernand Pouillon. The project aligns with the orthogonal grid of the surrounding urban fabric while maintaining an open, fluid, and evolving relationship with the ground plane.

At the project's core is a 600-seat auditorium around which the building unfolds. Unlike traditional performance halls, often enclosed in opaque volumes, the auditorium's envelope rises and opens to reveal the circulation spaces and interior functions. This design choice prevents the "black box" effect, instead integrating the performance space into the urban environment.

In contrast to Pouillon's repetitive modular windows, which define the surrounding residential buildings, this project introduces free-form bay windows, whose geometry echoes the branching patterns of trees. These irregular openings break the uniformity of the façade, offering dynamic relationships between interior and exterior.

Self-portrait Robert Doisneau, 1947

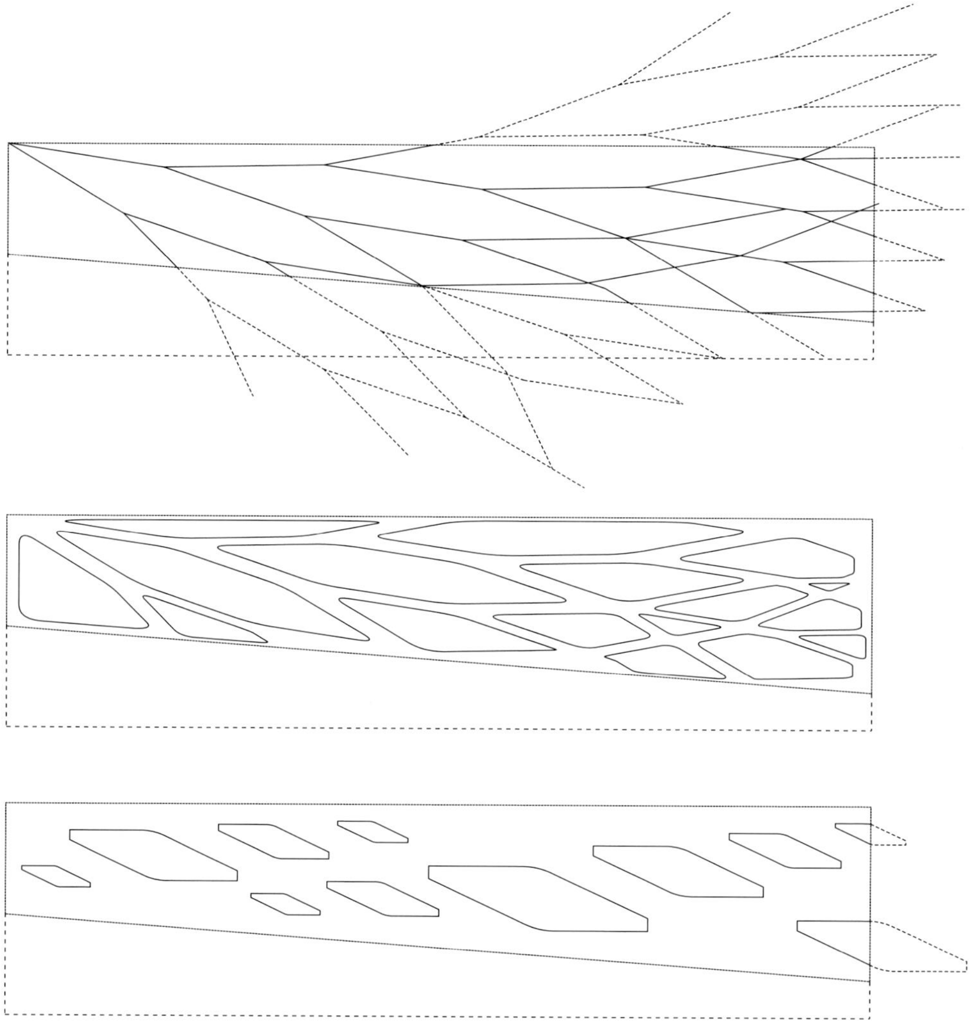

From branching geometry to leaf shape cells

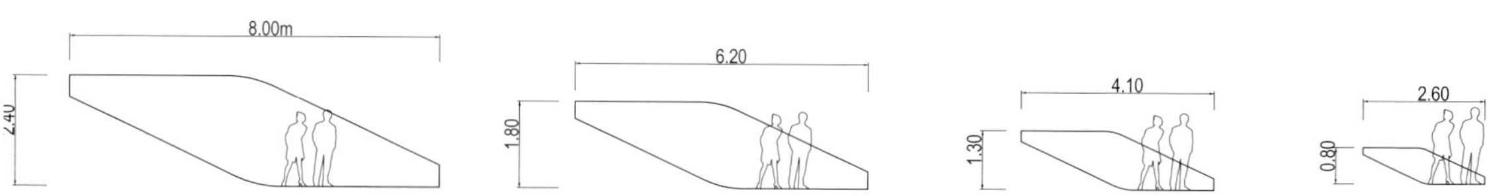

Four different scales of leaf shaped windows

The project is designed from the inside out, emphasizing fluidity between ground, roof, and built volume. The cantilevered structure extends over the reception hall and forecourt, creating broad, glass-enclosed spaces sheltered from sun and rain. Around the periphery, reception and meeting spaces, offices, and art studios unfold across two levels, fostering an open and interconnected environment.

All circulation and reception areas are flooded with natural light, either through the transparent curtain wall of the reception hall or via the shed roofs punctuating the workshops. These elements illuminate the interiors and shape the building's luminous atmosphere. The variation in size of these components creates a dappled lighting effect reminiscent of sunlight filtering through a forest canopy.

Inside the main hall, the light evolves throughout the day, subtly transforming the space. As dusk falls, integrated lamps within the sheds, equipped with photosensitive sensors, gradually take over, ensuring a seamless transition from natural to artificial illumination.
Meanwhile, offices and workspaces benefit from large, irregularly shaped bay windows, which frame dynamic views of the outdoors and reinforce the building's strong connection to its landscaped surroundings.

An Architecture that Breathes, Grows, and Evolves

This project transcends the conventional boundaries of performance halls by embracing the organic logic of nature. It is an architecture that interacts with light, climate, and surroundings, framing experiences rather than enclosing them. Like a tree, it stands rooted in its structurally efficient and spatially fluid environment, shaping a space where architecture and nature converge.

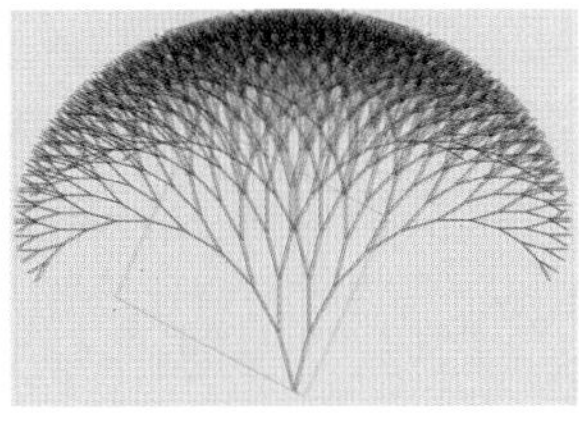
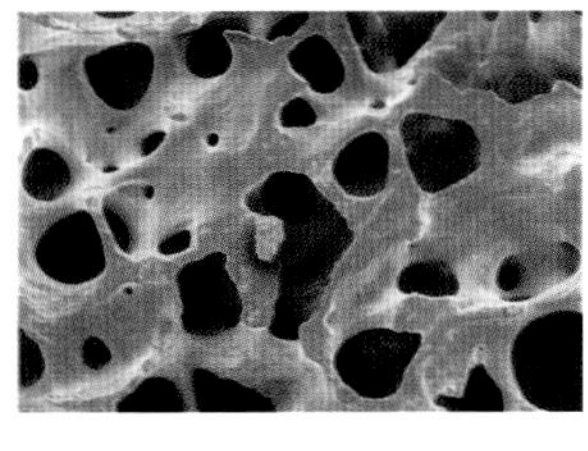

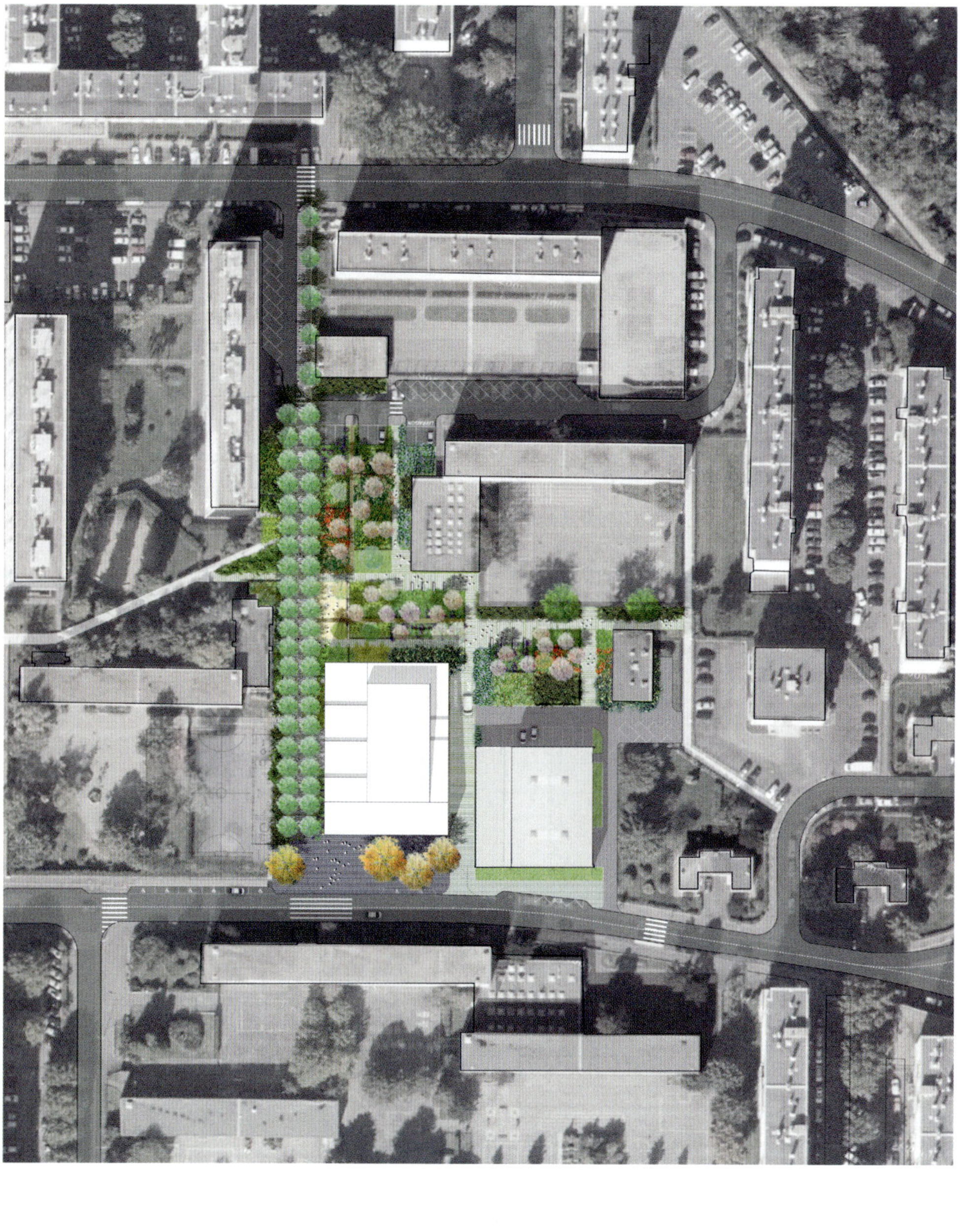

Art center site plan in F. Pouillon. Master plan for Meudon-la-Forêt

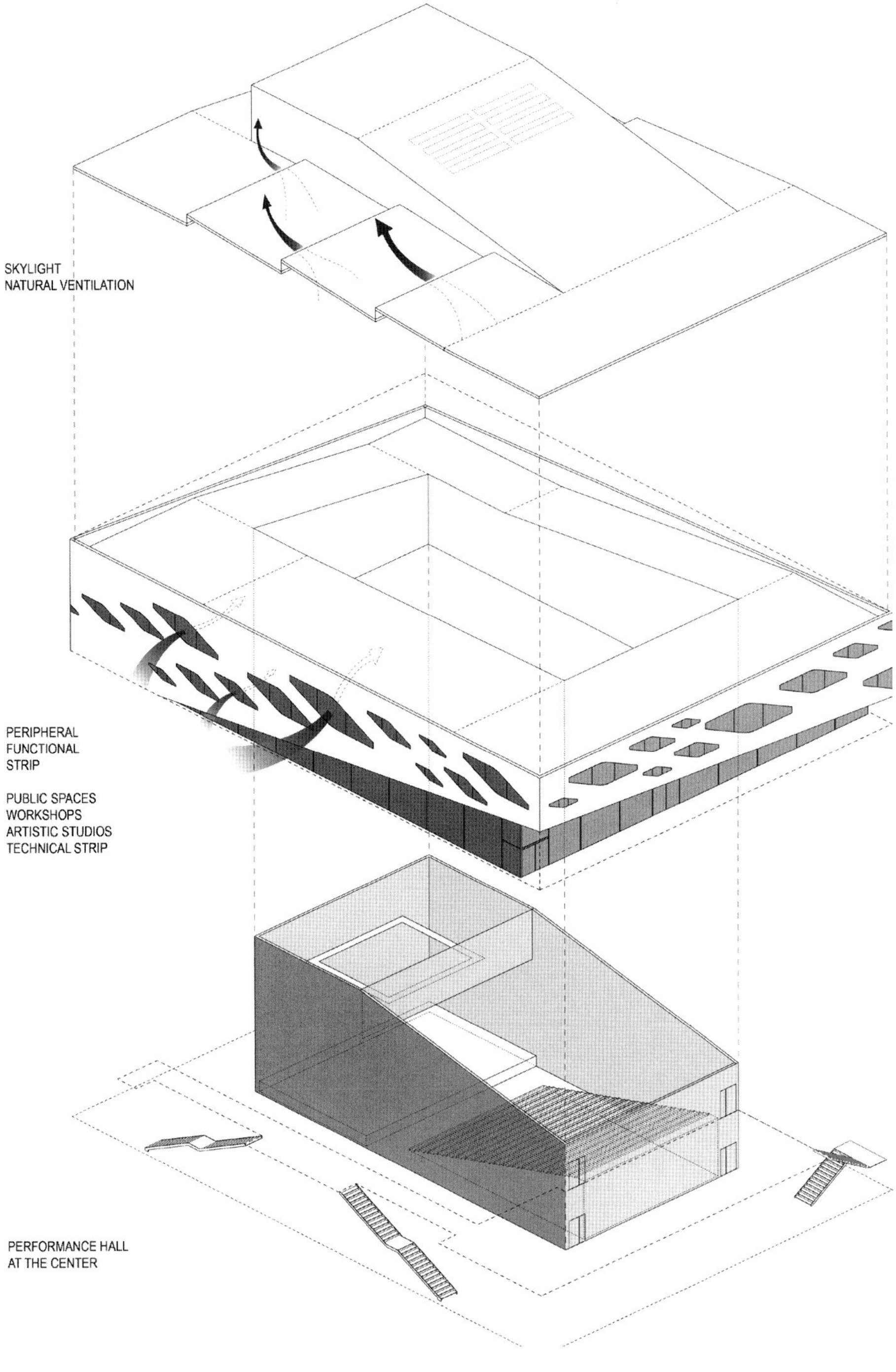

Central performance hall

Building edge with "window's leaves"

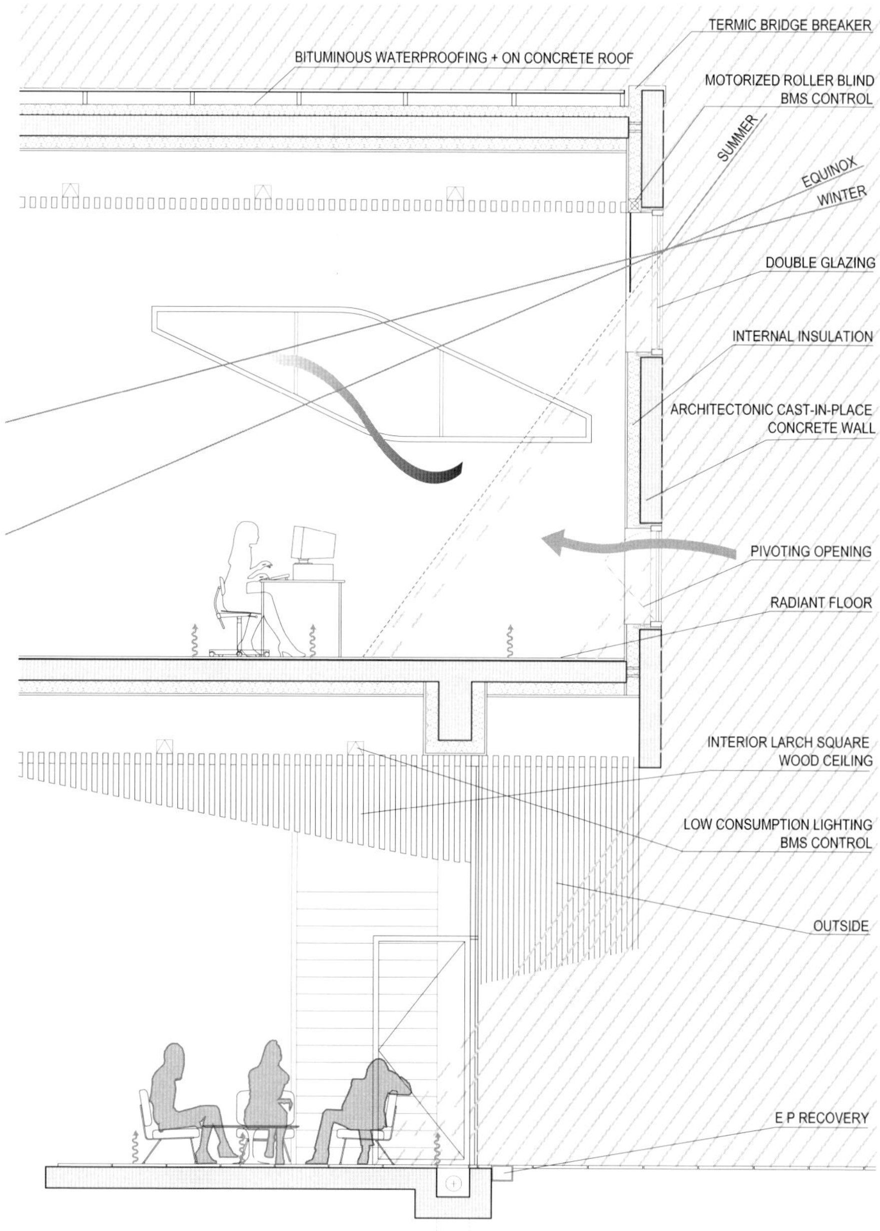

Detailed bioclimatic section of building

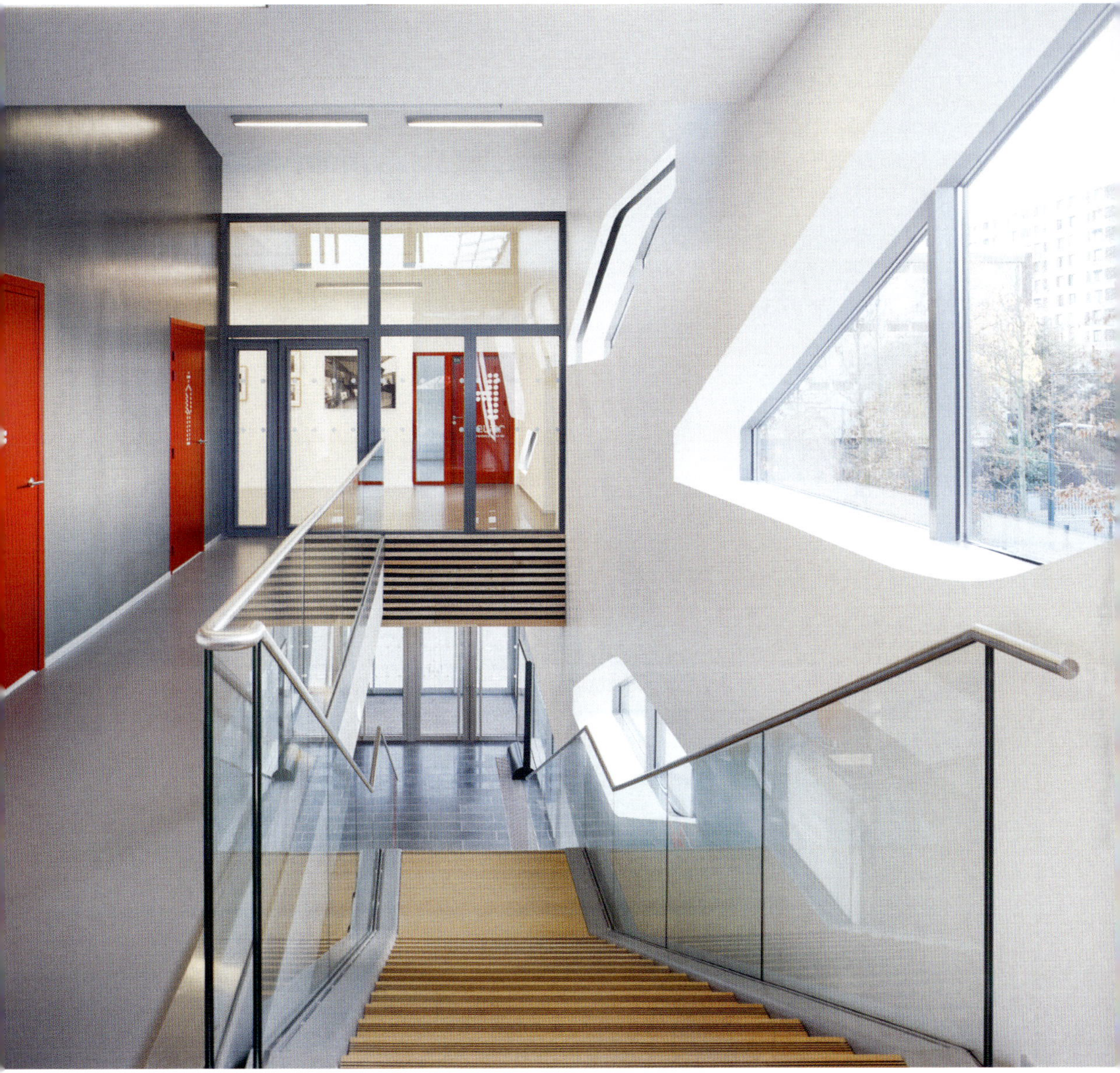

View of leaf window gallery

Interior view of concert hall

ESPACE
ROBERT

Belle

EIFFEL DNA

Paris, 2012

"... the authentic laws of the energy are not always conformal with the secret laws of the harmony..."
Gustave Eiffel

The Eiffel Tower, an enduring symbol of Paris, suffers from the weight of its success. Since its inauguration, the number of visitors reaching its summit has steadily increased, pushing the structure to its capacity limits. Today, 6.5 million visitors per year endure waiting times ranging from 35 minutes to over an hour to access its top floor. The tower's tapering geometry further exacerbates the issue, as the floor area decreases with height, creating congestion and crowd management challenges.

To celebrate the tower's 120th anniversary, we envisioned a speculative project that rethinks its public reception and access areas. The proposal introduces a temporary horizontal extension of the third floor, designed to enhance visitor circulation and the experience of the breathtaking 360-degree panorama of Paris.

When Gustave Eiffel designed the tower, he saw it as a platform for scientific exploration, capable of supporting various instruments for studying gravity and wind pressure. The structure was dimensioned to accommodate additional weight, a foresight that proved valuable during World War I when the tower housed antennas for national radio transmissions. Structural analyses confirm that the Eiffel Tower is a highly hyperstatic structure designed for loads greater than it currently supports.

By leveraging this structural resilience, our project proposes a lightweight carbon-Kevlar extension temporarily bolted onto the third-floor slab without altering the original framework. This intervention would expand the usable floor area from 280 m^2 to 580 m^2, significantly improving accessibility. The extension consists of two additional platforms, seamlessly integrated with the third-floor deck and the upper level—formerly Gustave Eiffel's private apartment—preserving the spirit of the tower's historical significance.

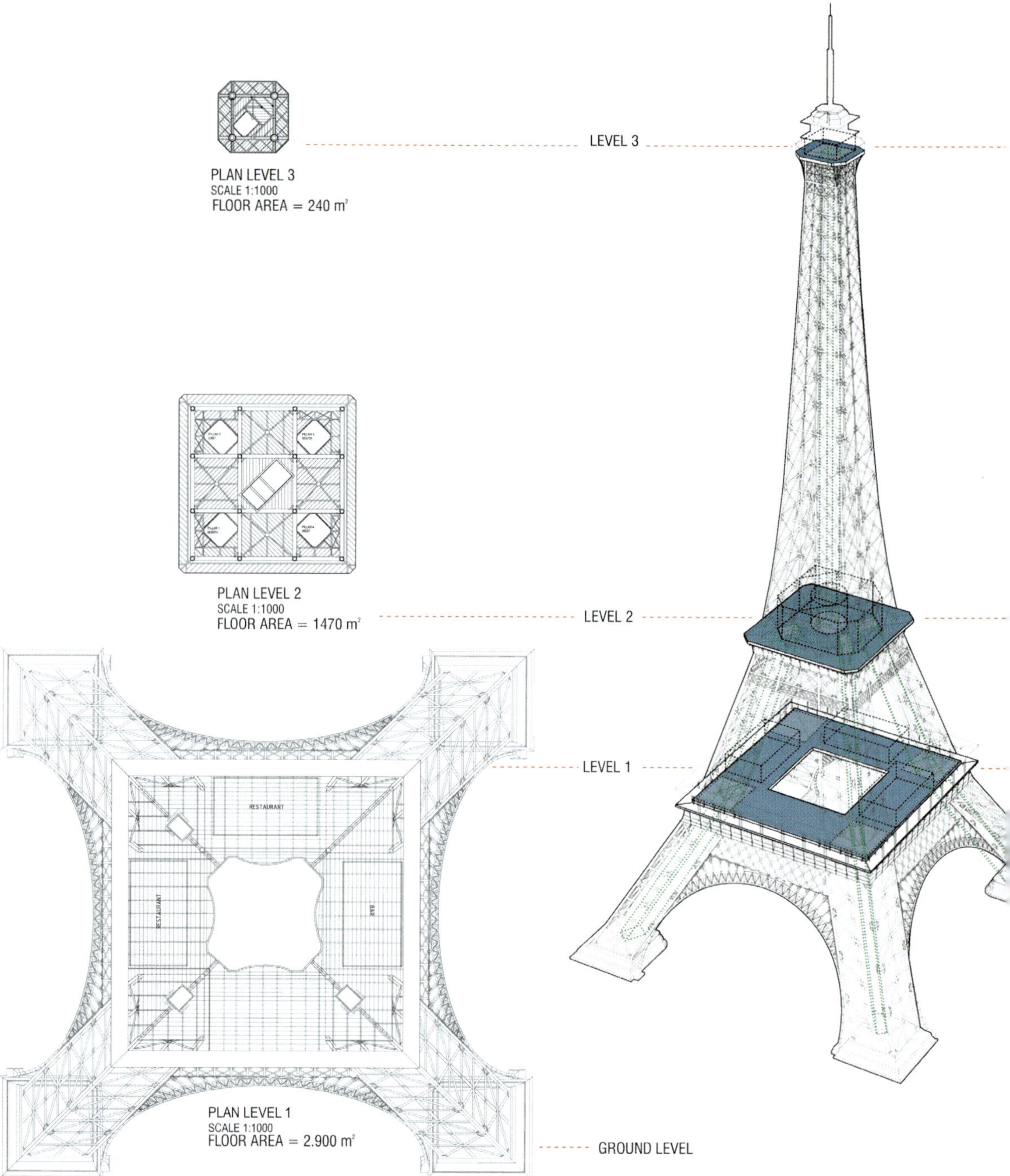
PLAN LEVEL 3
SCALE 1:1000
FLOOR AREA = 240 m²
LEVEL 3
PLAN LEVEL 2
SCALE 1:1000
FLOOR AREA = 1470 m²
LEVEL 2
LEVEL 1
RESTAURANT
RESTAURANT
BAR
PLAN LEVEL 1
SCALE 1:1000
FLOOR AREA = 2.900 m²
GROUND LEVEL

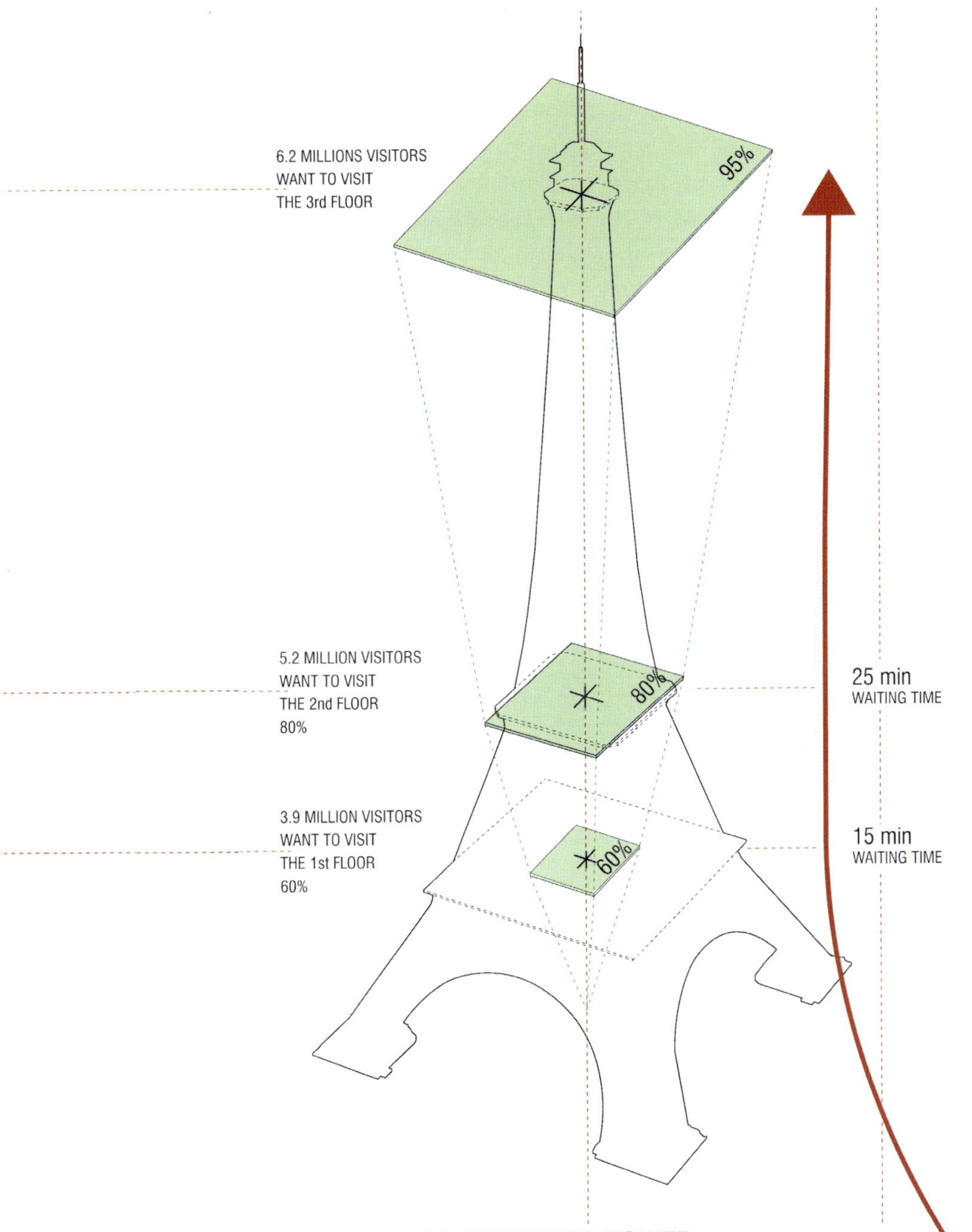

6.5 MILLION VISITORS VISIT THE EIFFEL TOWER 95% OF THEM WANT TO VISIT THE TOP FLOOR

The design is parametric, inspired by Eiffel's structural principles of three-dimensional cross-bracing beams. Using generative scripting, we developed a branching system that emerges from the tower's existing framework, echoing its distinctive lattice DNA. The script, anchored within the 10 x 10-meter cube at the summit, generates three interwoven structural layers, forming a complex yet highly efficient weave. This non-repetitive pattern enhances structural performance, diverging from conventional modern engineering, which relies on repetition and optimization. Instead, this approach explores an alternative model of high-performance design, where organic variation strengthens the overall system.

By temporarily extending the Eiffel Tower's upper levels, this project redefines its capacity and reinterprets its legacy, bridging historical ingenuity with contemporary architectural exploration.

Branching patterns typology for 3rd-floor growth

Planar study of extension

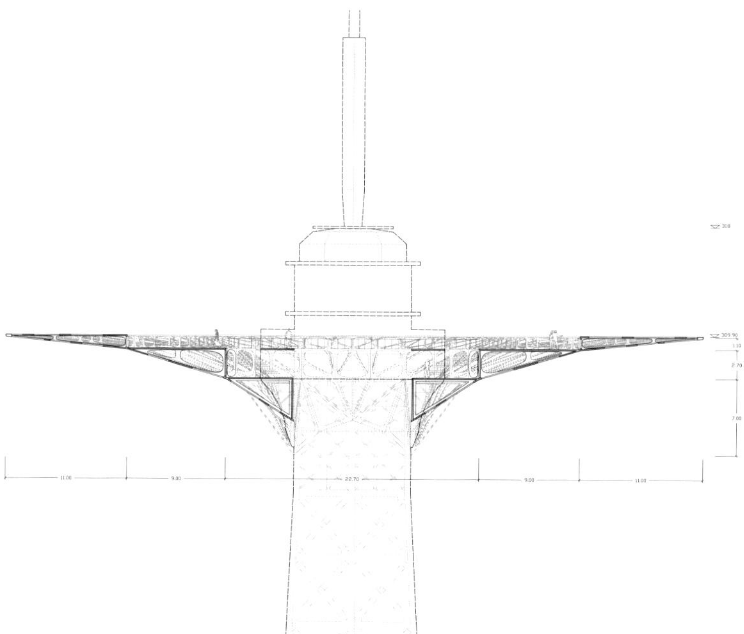

Head structural reinforcement section

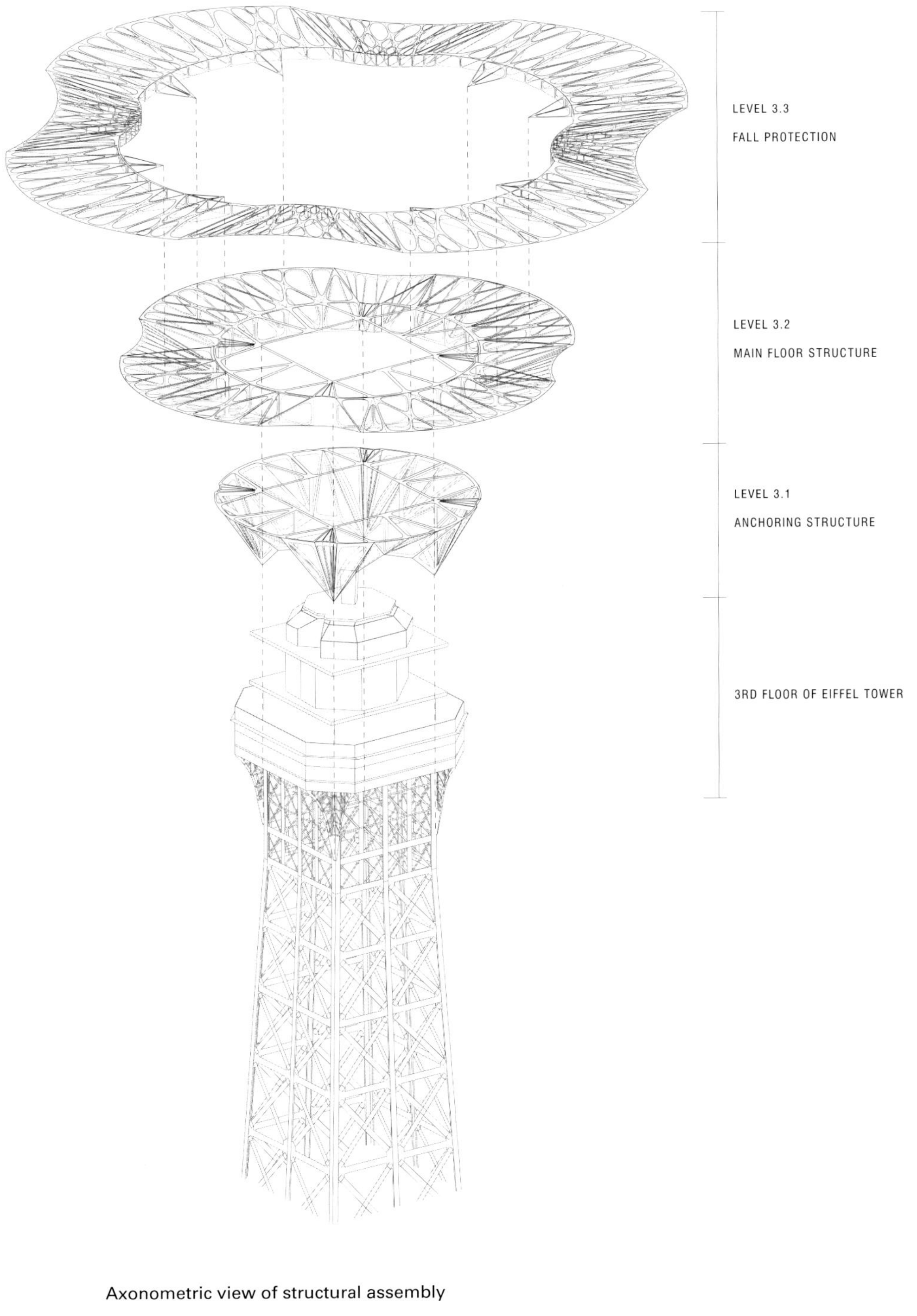

Axonometric view of structural assembly

MODULATIONS

Jardin de Métis / Reford Gardens
Grand Metis, Québec 2008

Architecture has long been considered the creation of a geometric and, therefore, artificial order in opposition to the garden as a natural order. Today, this distinction is increasingly blurred at a time when architectural practices are inspired by natural processes, evolutionary models, and modes of transformation. It is also evident that architecture has long defined a framework for the garden. Beyond the debates of form and function, it is a broader practice that defines a system of ambiances, a continuous "climate" encompassing the territorial, landscape, urban, architectural, and bodily scales.

The practice of landscape has long been based on staging nature through regulating and orthonormal principles of axes and lines. The natural elements were calibrated and "celebrated" a known relationship of man to his environment. Nowadays, the city and the garden are less and less distinct entities. The practice of landscape promotes the emergence of new environments through the development of gradual ecological systems and the progressive transformation of contextual and specific elements.

The garden, a metaphor for an original paradise and place of re-creation of nature, has become a specific field of creation that feeds on conceptual methods developed, among other things, by artistic and architectural influences. It maintains a more ambiguous relationship between the architectural horizon and the natural environment. This almost instantaneous work on space is thus defined in several ways.

Fern leaf growth photographies

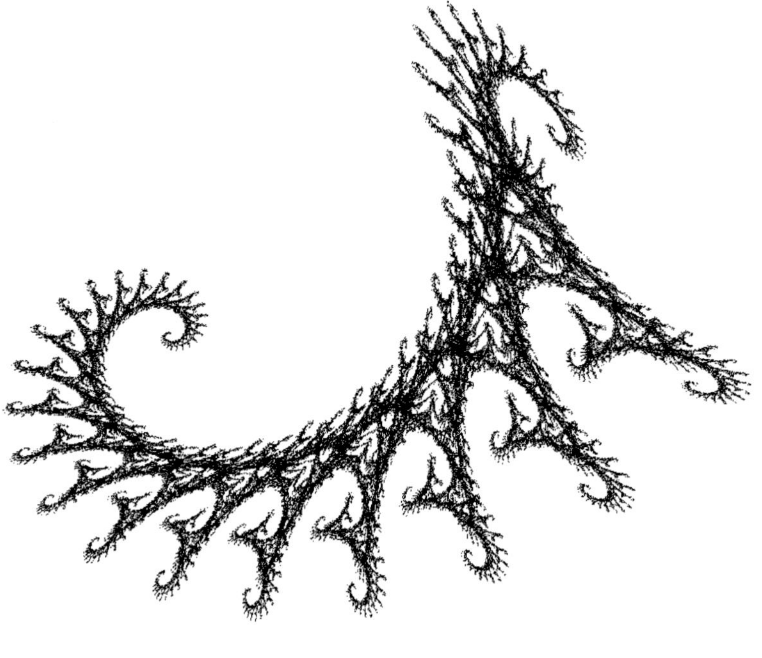

Algorithmic fern like geometry

Initial steel weawing prototypes

The temporary garden is a contemporary mode of expression of space based on mass consumption, perishable in time (an annual change), but where the concepts can be adapted elsewhere. It is not the specific expression of a medium but participates in a global network of exchanges between different creators.

The garden conveys an idea of nature where "raw elements" are forced to dialogue in a restricted space. The garden is defined by the relationships it establishes, a relationship of friction between natural and artificial elements*. It is a laboratory space, a place of experimentation.

The intention is not to seek harmony or homogeneity but a dialogue, a new articulation, where each visitor experiences his way of crossing the space. Thus, our garden only takes on meaning and can only be understood when it is put into "action" by the visitor.

We explored this at Grand Métis in 2008 when creating the temporary garden "Modulations" with the landscape designer Philippe Coignet. Inspired by the simplicity of a fern leaf, we deployed its mode of growth and geometric formation on a larger scale to create stainless steel screens. These large strips of metal were developed by simulating plant growth, with plant growth and foliage generation parameters analyzed for their ability to reflect and conceal surrounding space. The geometry of the steel strips reproduces that of the sheet itself, reflects and diffracts the light of this space in an almost mimetic way to the plant.

The complexity of the natural phenomena involved in the landscape is now better understood. Biological and computer research makes it possible to understand the rules and processes of natural formation, to divert them and to draw inspiration from them in order to create new forms.

The vegetal motif is the pretext for the invention of a new garden, both simulated reality and evanescent surface, which forces the visitor to move and discover. This project engages a reversal of the limits of the garden inside itself by a visual device. Only during the movement of our body in the garden do the reflections and the play of light on this surface make this space in transformation readable. These changes then express the notion of temporality.

LEARNING AND RESEARCH CENTER

Issy-les-Moulineaux, 2017

The restructuring of the CNFPT training center in Issy-les-Moulineaux aims to create an open and dynamic space for educational innovation. The project includes 22 training rooms, lecture theaters, digital workspaces, co-design areas, and offices. At its core, the intervention focuses on opening up the courtyard by removing multiple concrete load-bearing walls, allowing natural light to penetrate the heart of the building and introducing alternative learning spaces.

We approached the building as a porous structure, carving out new free and 'informal' spaces that complement the traditional training areas. This interplay between the external envelope and the interior layout maximizes daylight diffusion, creating luminous 'pockets' within the core of the building. The designers enhance these bright zones by using a refined approach to acoustic partitions that vary between transparency and semi-transparency, fostering visual and spatial continuity.

The reconfiguration eliminates conventional corridors, replacing them with a vast three-level "third-place" space. This multifunctional environment is designed not only for reception, interaction, and self-learning but also for flexibility, fostering a new model of innovative education.

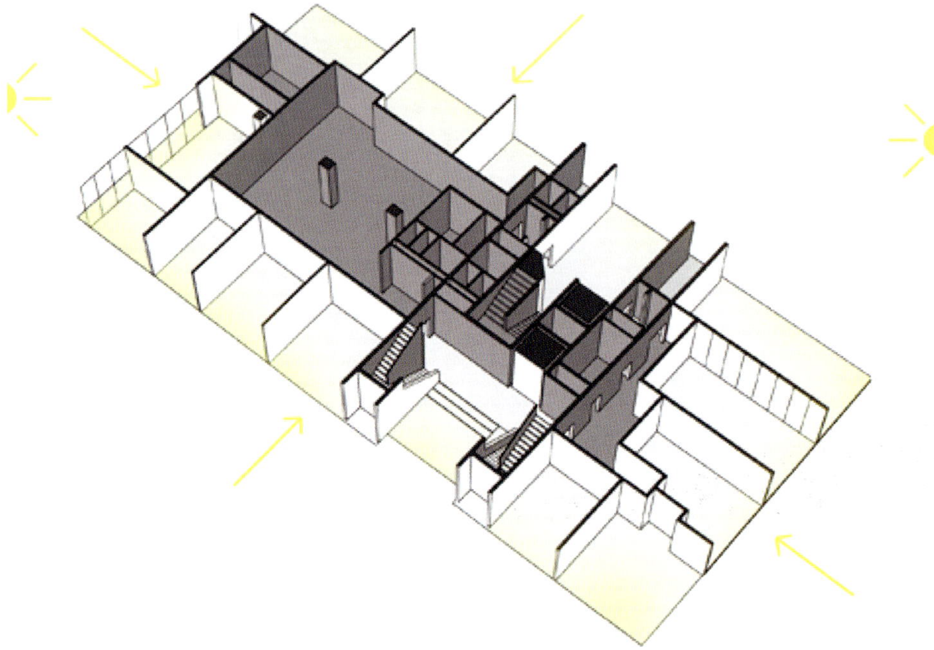

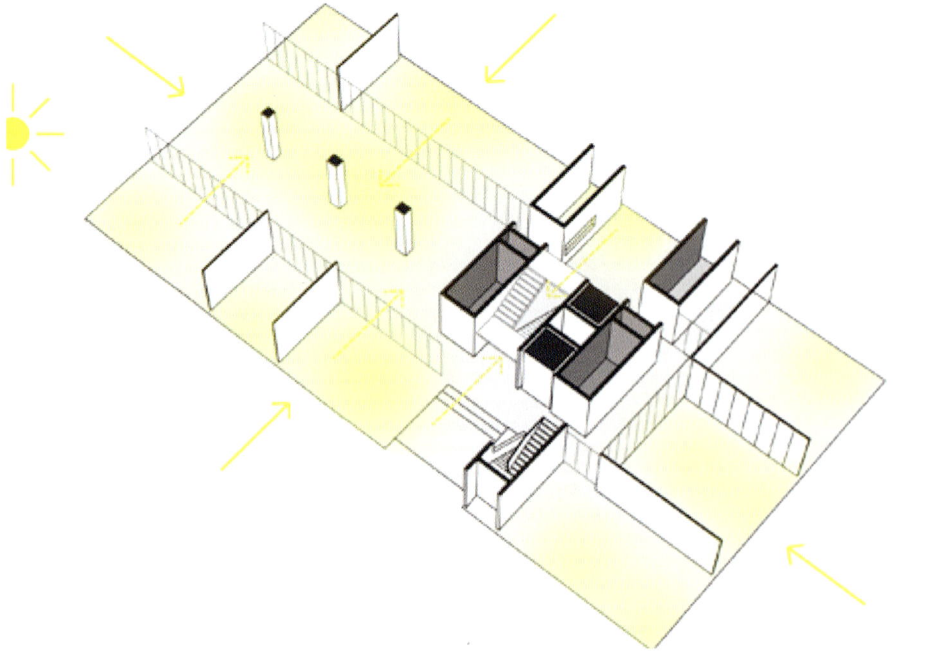

Opening floor plan education rooms
for natural light flow

This central space serves as a learning hub on the ground floor, housing a documentary collection and a media library. The first floor transforms into a convivial space, integrating a cafeteria and a co-working area. In contrast, the second floor hosts a collaborative workshop that extends onto a vegetable garden and a rooftop terrace. Here, residual spaces—corridors and staircases—typically become active contributors to knowledge production and exchange.

Rather than imposing a new aesthetic onto the existing building, the design approach was careful deconstruction. By stripping back successive layers, we revealed the raw essence of the structure—prefabricated concrete slabs, columns, ventilation ducts, and cable trays—exposing them as integral elements of the user experience. This process of unveiling the building's inner framework allows for a deeper engagement with its materiality and spatial identity.

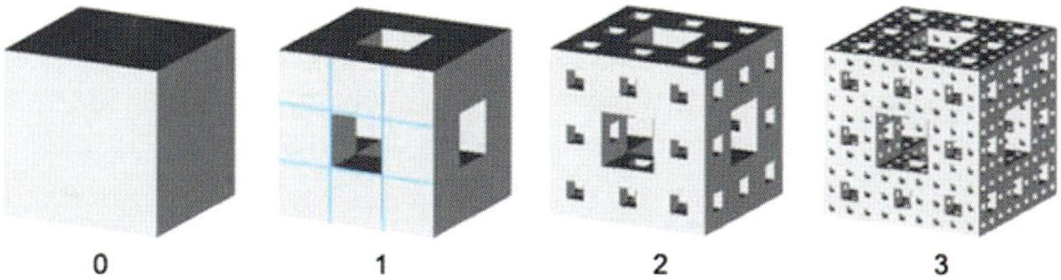

Sierpinsky sponge geometric model used for the building's internal carving out

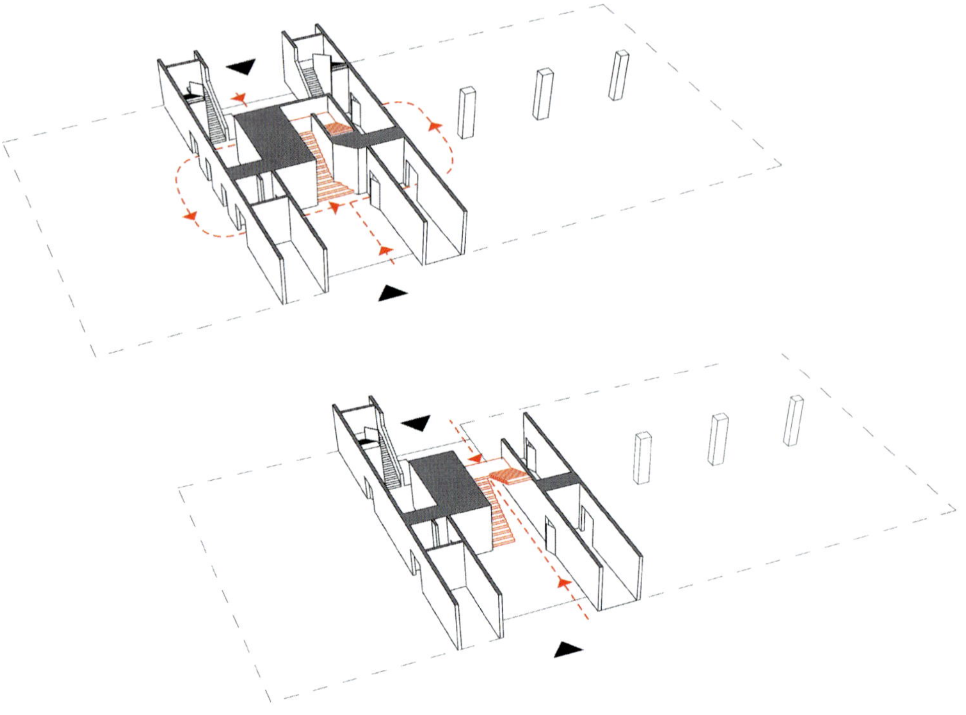

Staircase opening process

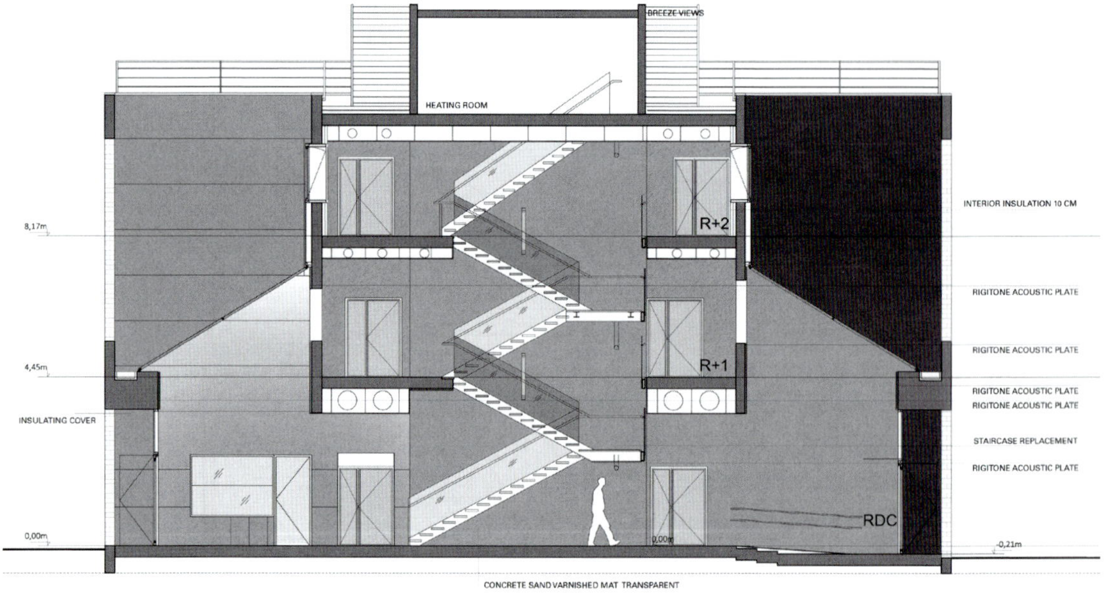

Main staircase sectional study

CFA

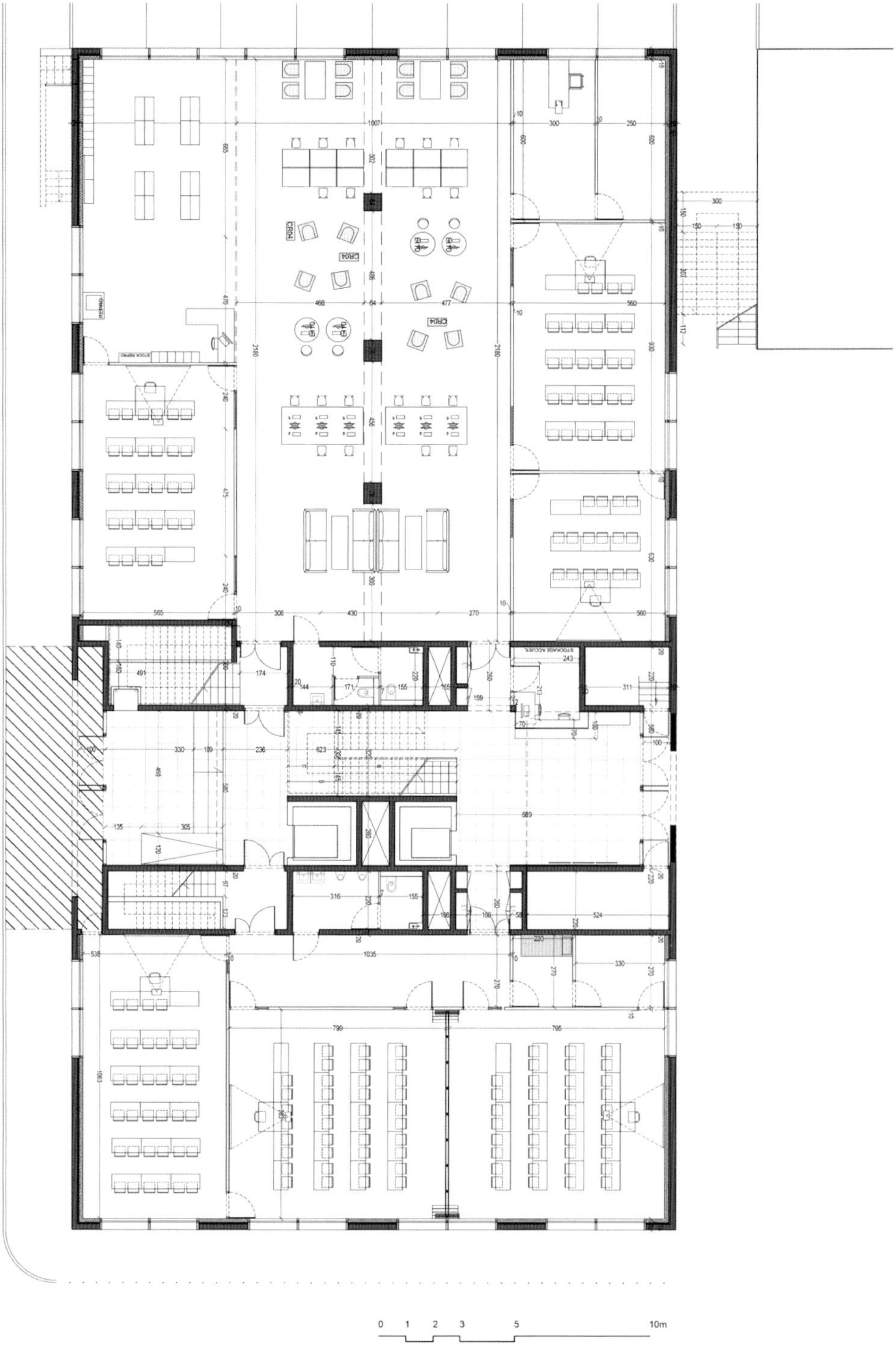

Ground floor plan

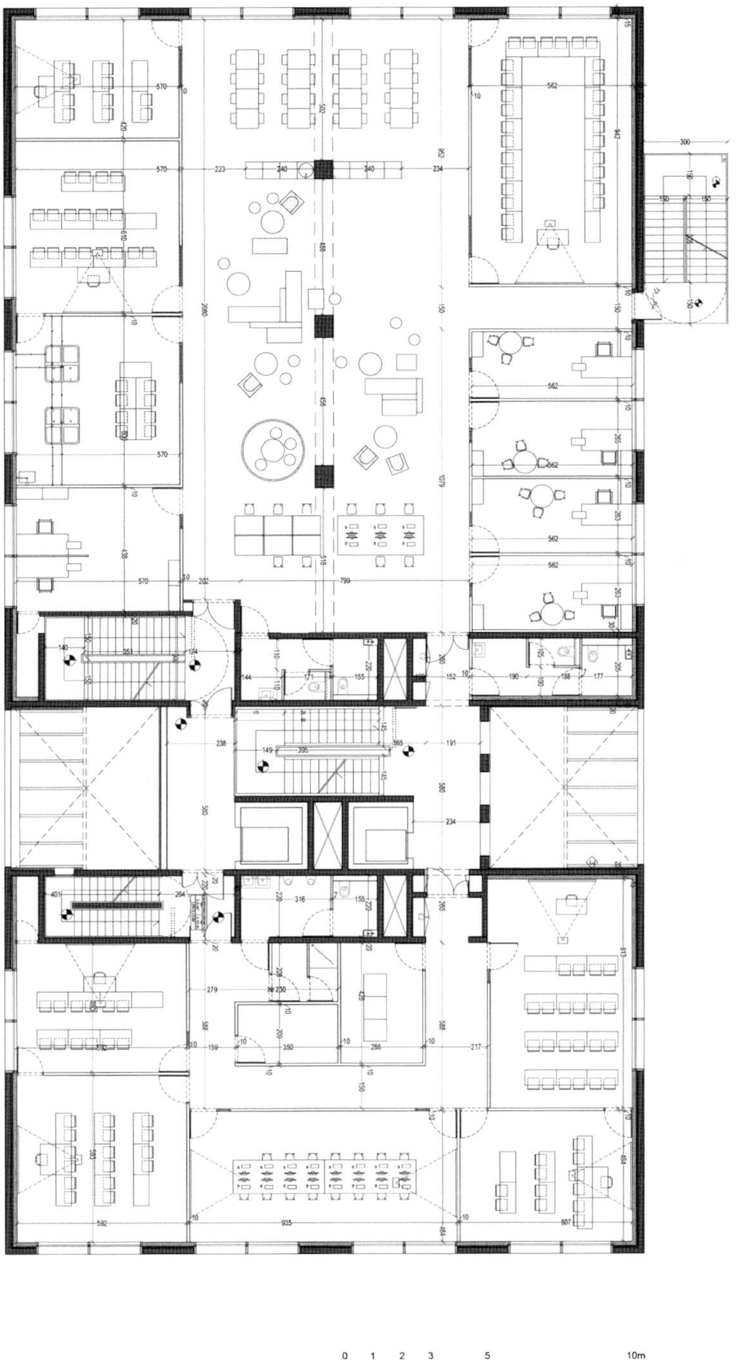

First floor plan

Steel and glass "hub" staircase

Chapter 2

DATA ENERGY MATTER

Energy

Implementing Low-Carbon Architectural Practices

Our philosophy asserts that energy and architecture are inherently intertwined, forming a unified challenge that shapes both design and sustainability.

Whether in the project design process, in the choice of construction method, in the life cycle of materials, or in the choice of the building's energy system, the architect's work is also that of an energy specialist. He designs and develops systems for the consumption, production, and renewal of energy, whether technical and physical, human or material. In each of our projects, we have sought a dialectic within the architectural project between bioclimatic devices, derived from the study of original, rural, or vernacular architectures, with more advanced technologies of adaptation or optimization allowing interaction between humans and the spaces of our projects. (Adaptive architectural envelopes, routines, and algorithms are linked to uses, as well as adaptation of natural ventilation and solar control).

With the implementation of new European objectives for energy reduction and city optimization, low-carbon architecture practices are finally gaining momentum. The project management teams are developing a comprehensive approach to the overall environmental impact and life cycle of a building: bioclimatic architecture, optimizing the use of natural resources and the implementation of new bio- and geo-sourced circuits, designing the building in relation to a shared or autonomous energy network.

The first three projects presented here, the library and auditorium of the Amiens University Campus, the Bayeux media library and the "Climat Planète" research center, are three significant achievements in the development of this approach.

The Amiens project has made it possible to develop an architectural envelope system for solar control using scales with variable angles—with seasonal adjustment of their orientation—based on a bio-inspired model such as that of the pine cone. The Bayeux media library is the result of a project in which the windows and curtain walls, the interior filtering screens, and the roof overhangs are precisely calculated to magnify the transparency of the building with optimal solar and thermal protection. The "Climat Planète" project, a climate research center in Grenoble, aimed to find a perfect energy balance between groundwater temperature, cooling of a data center, and thermal regulation of researchers' offices and laboratories.

In the case of the Hellenikon Metropolitan Park in Athens, a project to restructure the huge site of the former Athens airport, the team's reflections were inspired by the natural processes of space recolonization to be applied to an airport infrastructure while limiting the energy used and the cost of the project. We were particularly interested in water management on the site and in how large-scale landscaping could channel the force of frequent flooding in this area due to soil artificialization through topographical profiling adapted to the collection and storage of runoff water and initiate a new virtuous water cycle.

The latest project for the University of Orléans presents a unique approach to the energy renovation of buildings, a global project to transform and adapt buildings to climate change that would affect more than 60% of the world's building stock. Here, we have combined a project for the educational transformation of lecture halls through cluster working systems, with external thermal insulation with concrete resin cladding, combined with an original approach to welcoming biodiversity. In partnership with the Natural History Museum of Bourges, we identified the presence of a local species of bat on the building, the common pipistrelle, which we were able to relocate after the work by means of a system of nesting boxes integrated into the façade and its insulation.

A Simulation Model to Rethink the Energy Isuue

Could we propose a new approach to the architectural project by combining the tools for analyzing and calculating physical phenomena with the expertise of an architect in an urban environment?

The essay presented here is the result of research, halfway between architecture and engineering, and is based on a digital simulation model, which allows an analysis of the behavior of the building in the design phase. This information constitutes a project design environment aimed at "profiling" the building, according to an analytical and performative method rather than a simple subjective approach to architectural design.

The work and research are based on a dual body of knowledge: architectural design and fluid mechanics, making it possible to bring architecture out of its isolation between the arts and engineering and to propose a cross-fertilization of these disciplines. The approach is, therefore, transdisciplinary and aims to study the impact of wind in the hyper-localized areas of large buildings and to inform the architectural design of the inevitable change in airflow at the periphery of its envelope.

This analysis reveals that every building, every wall, generates by its presence on a plot of land, an air pressure differential on its external walls, linked to the displacement of the surrounding air layers. This study has shown that the volumetry and position of a building in its context can thus be "profiled" and "oriented" to increase and control the wind speed at its periphery.

This method of designing buildings for architects and engineers would allow the integration of energy capture devices such as micro wind turbines into the envelopes of their buildings. Aerofoil-type structures, such as lifting wings, can locally stabilize wind turbulence and selectively increase wind speed.

The improvement of the bioclimatic design of buildings brought about by this research, combined with global environmental policies to reduce greenhouse gases, is enabling the emergence of an ecosystem of energy sharing.

The emergence of the Internet of Things (IoT) allows the different components of the building to interact with each other, as well as with users and building management systems. This is an emerging infrastructure, the Internet of Energy, which aims to provide an organization for the sharing and exchange of energy at the local level. This new paradigm projects a vision of intelligent urban environments where energy resources, production, and consumption would be hyper-localized, even autonomous, with a virtuous environmental impact.

SCALES OF LIGHT

Auditorium and Library
Amiens, 2016

Located at the heart of the University of Amiens campus, a dynamic center of student life, the new library and auditorium form a key hub for teaching, research, and interdisciplinary exchange. The project has been designed as a natural extension of the landscape. It integrates harmoniously with the site's topography, creating a dialogue between the built space and the surrounding environment.

The design is structured around two platforms, both aligned with the university's central garden: the library to the south and the auditorium to the north, whose tiered seating follows the terrain's natural slope. Between these two spaces, a lobby is the main point of convergence, providing access to both volumes. A staircase, discreetly positioned in the gap between the two buildings, connects the lobby directly to the lower level, reinforcing fluid movement between different areas.

Natural light plays a fundamental role in shaping the library's atmosphere. Thanks to north-facing shed roofs that span the entire building length, the space at the heart of the structure is bathed in soft, uniform daylight. This overhead illumination creates an open and inviting environment, transforming the library into a place of exchange, knowledge-sharing, and interdisciplinary encounters.

The auditorium, built with light concrete walls, benefits from tall, continuous windows throughout, which allow natural light to permeate the space. These windows can be fully darkened using integrated roller blinds, ensuring optimal lecture, screening, and presentation conditions.

Designed on a free platform with minimal structural interruptions, the library's open-plan layout is supported by slender intermediate columns spanning 10.80 meters. A modular grid system of 1.80 m and 7.20 m organizes the structural framework, lighting, ventilation, and furniture distribution, creating an intuitive, legible, and highly flexible interior. This clarity of organization enhances the user experience, facilitating intuitive access to books and resources while balancing communal and quiet study areas. The transition from large reception spaces to smaller, more intimate work zones fosters an atmosphere conducive to both collaboration and individual concentration.

The building's façade functions as an intelligent skin, actively managing light, views, and climate control. Inspired by the scales of a pine cone, which open and close in response to humidity, the wooden sunshades adapt to the building's orientation, creating a dynamic interplay between transparency and protection. These vertical elements gradually shift in angle along the façade, adjusting to the spatial and environmental needs of the interior. Positioned to optimize western solar exposure, they offer superior filtering of low-angle sunlight, ensuring both thermal comfort and visual clarity.

Through its topographic integration, daylight strategy, and bio-inspired façade, the Amiens Library and Auditorium redefine the relationship between architecture, landscape, and knowledge, offering a harmonious and sustainable space for learning and exchange.

Library and auditorium site plan on campus

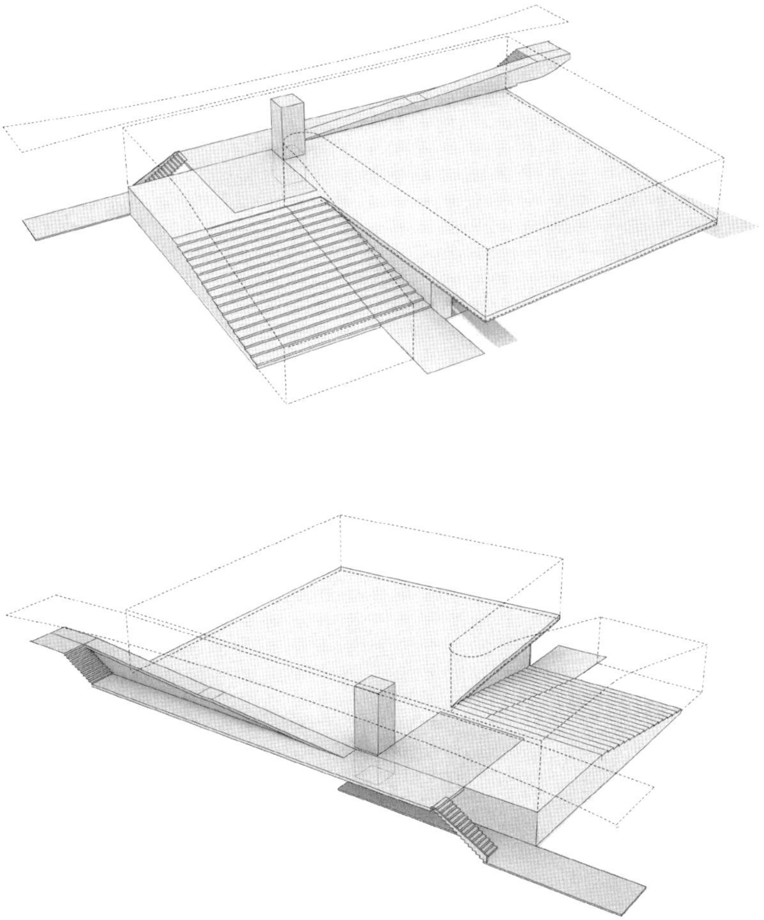

Circulation and volumetric diagram of project

AMPHITHEATRE

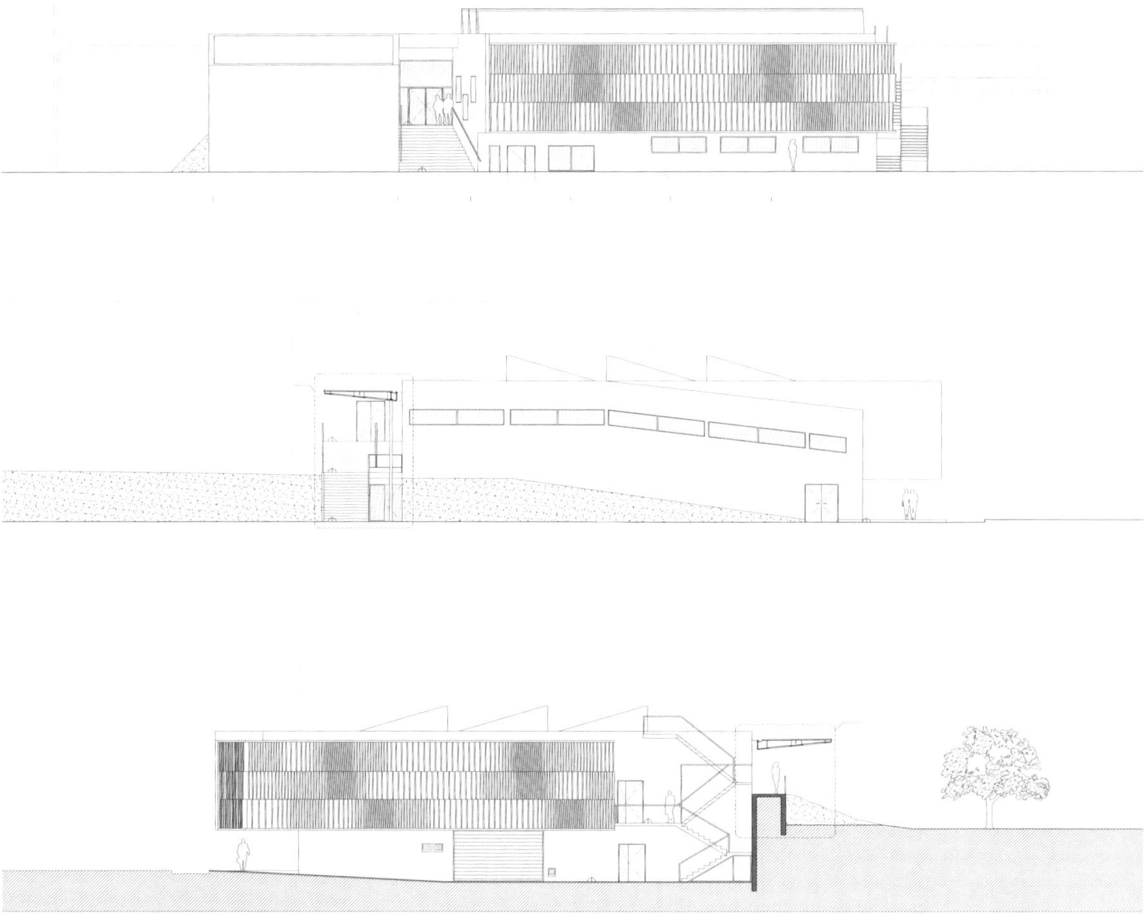

South elevation (library), east elevation (auditorium)
and west elevation (access gallery)

Pine cone scales as the biomimetic inspiration for reading hall shutters

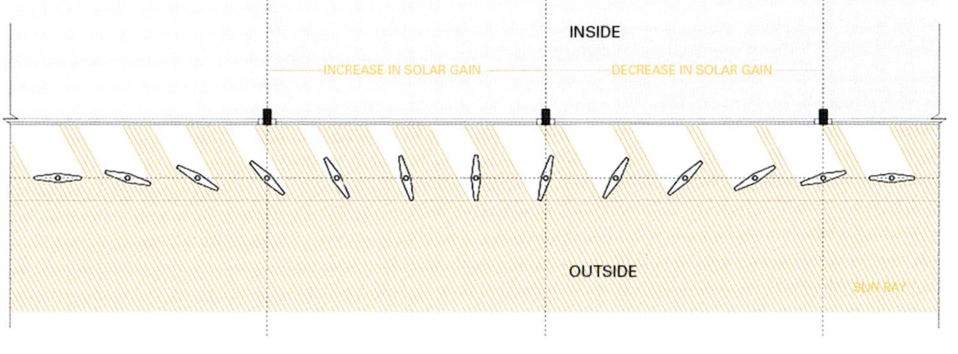

Shutters orientation study for facade design

Ground floor plan

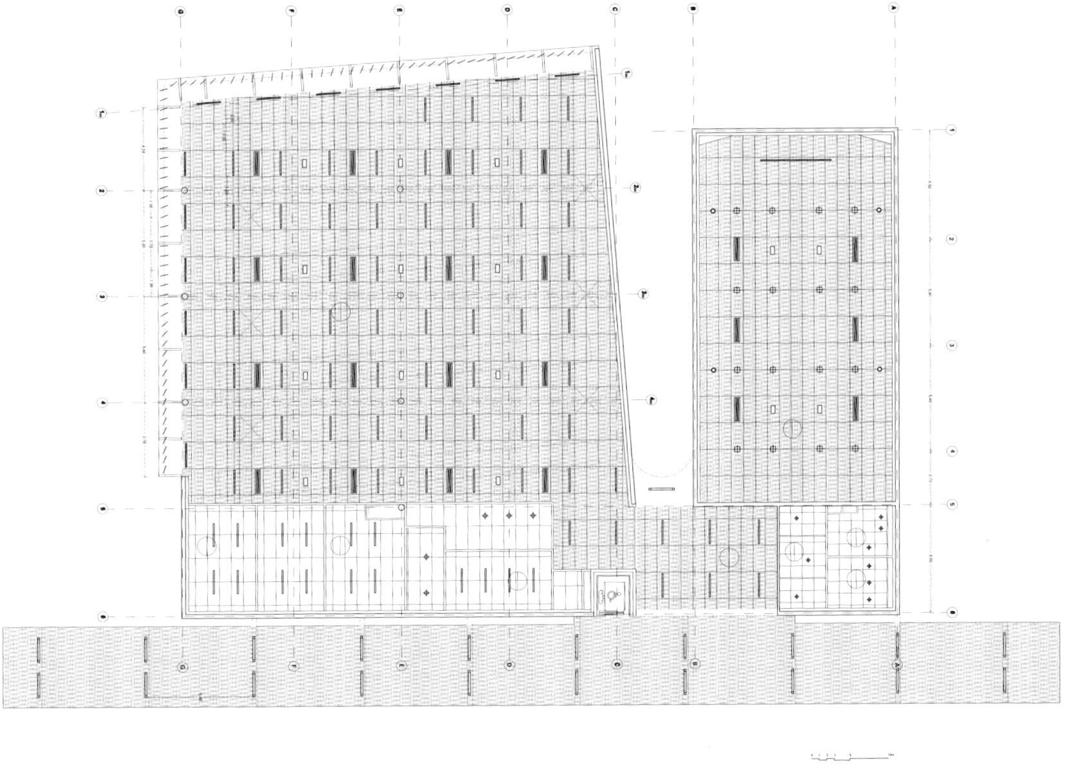

Ceiling plan for natural and artificial lights

AMPHITHEATRE

Connecting gallery from old to new campus

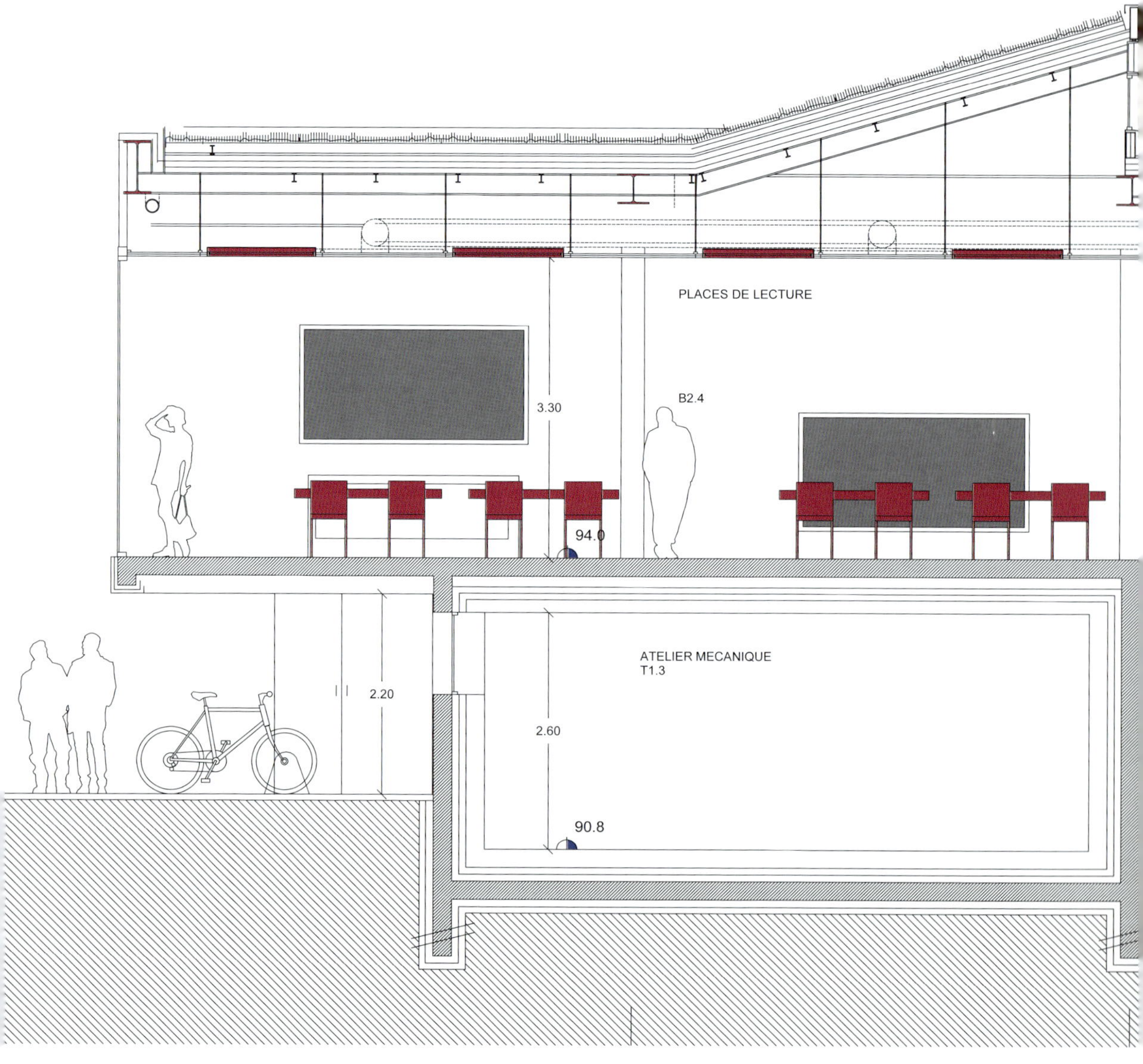

Sectional details of library facade and roof design

Amphitheater view

Natural light through wooden slats ceiling

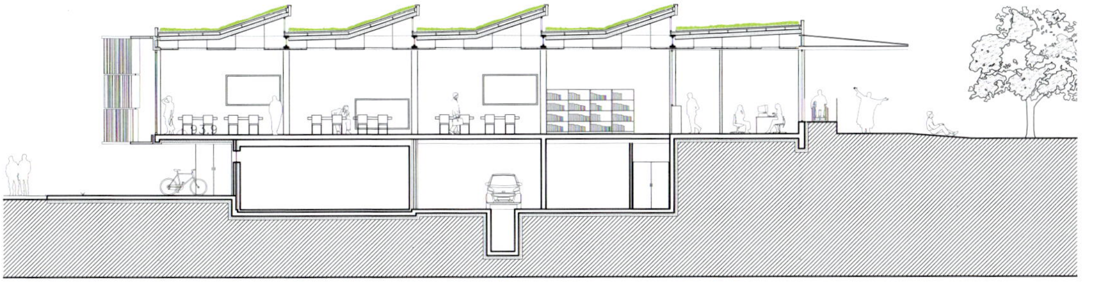

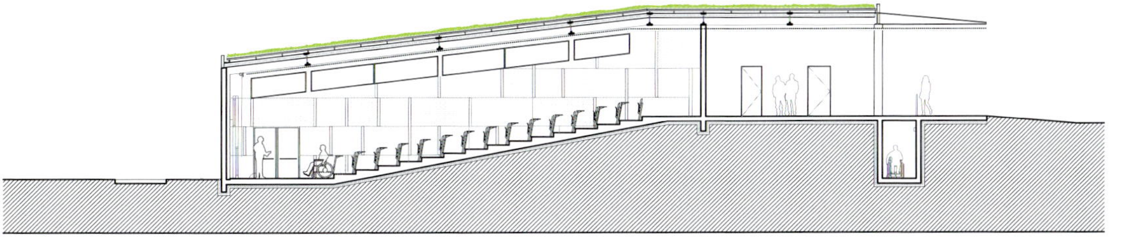

Library section and natural slope amphitheater

Following page: In between gallery space between light shutters and full glass walls

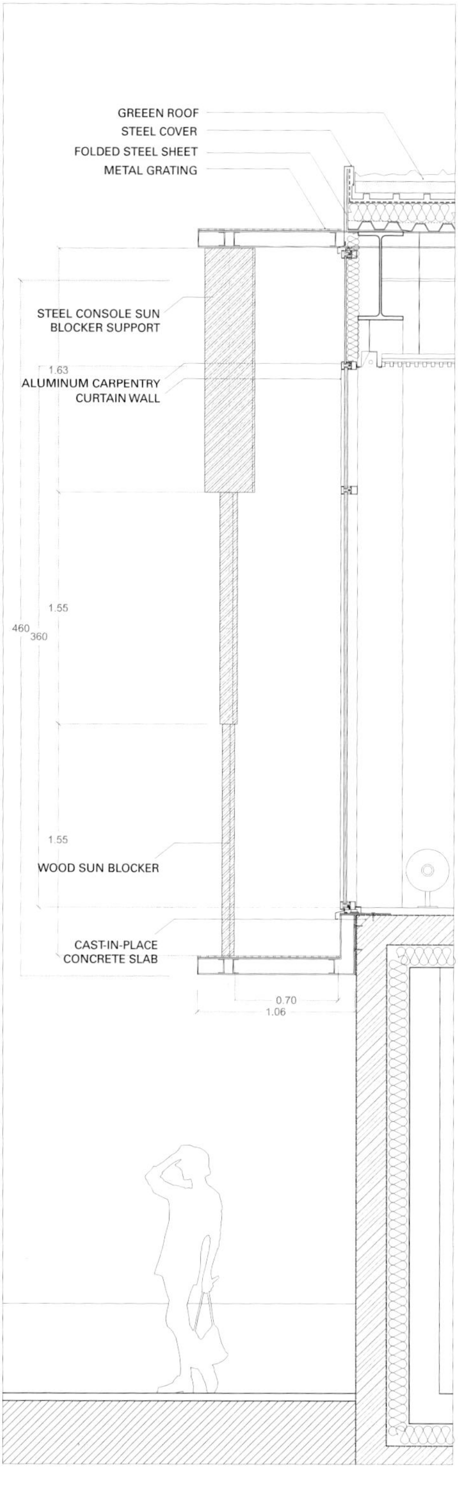

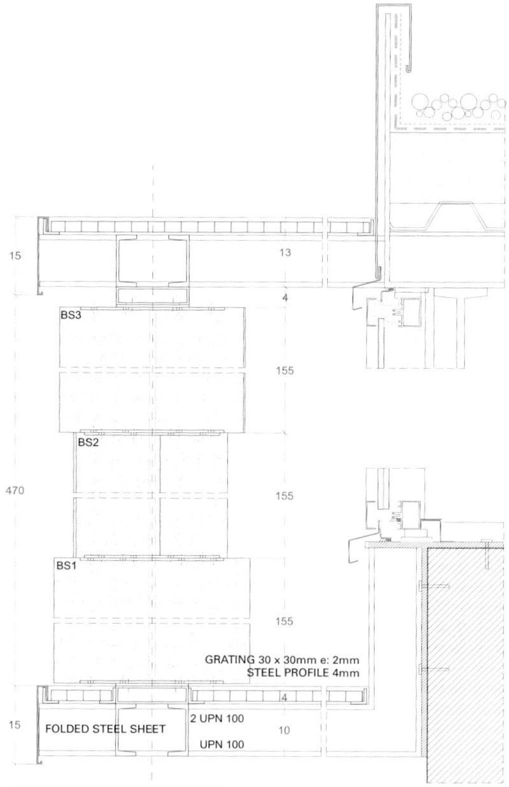

VERTICAL SECTION DETAIL 1/10°

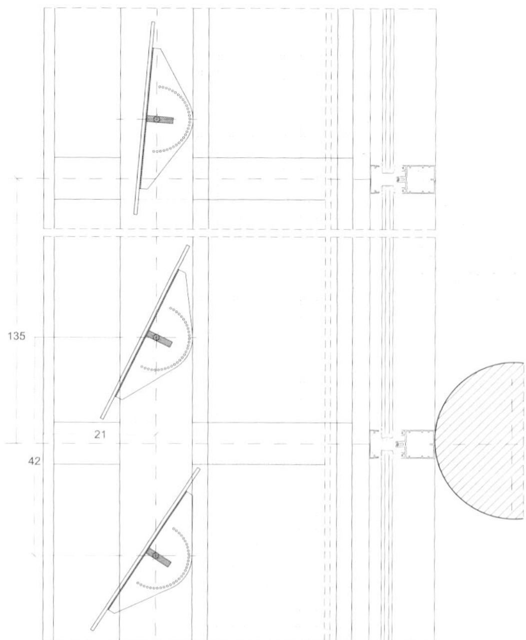

DETAIL PLAN 1/10°

Facade detail on shutter and technical gallery

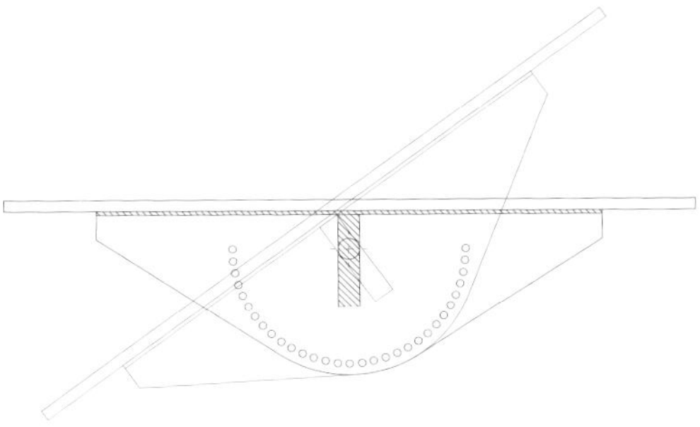

DETAIL OVERLAYING TWO BLADES BS 1/5°

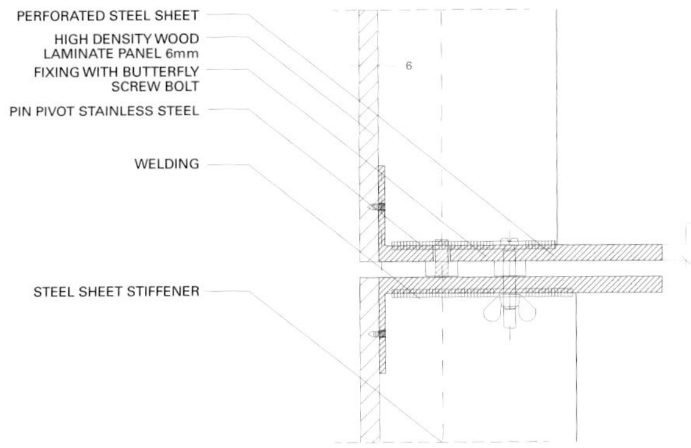

DETAIL VERTICAL SECTION
BS 1/1° FIXING

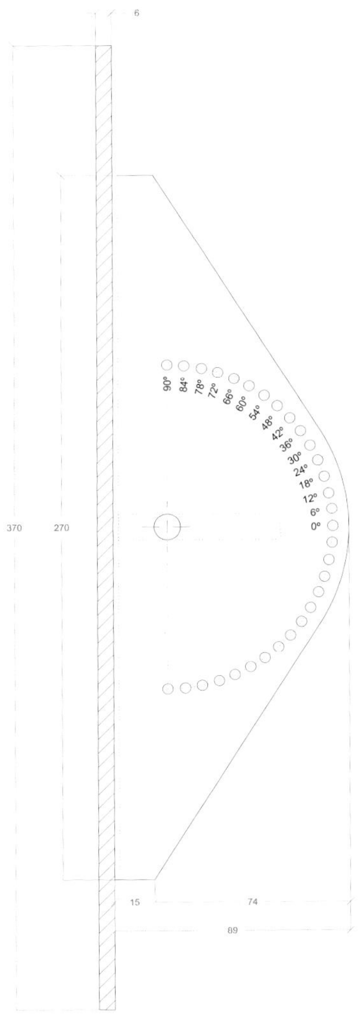

DETAIL SUPERIOR VIEW HORIZONTAL 1/1°. Dimensions in mm

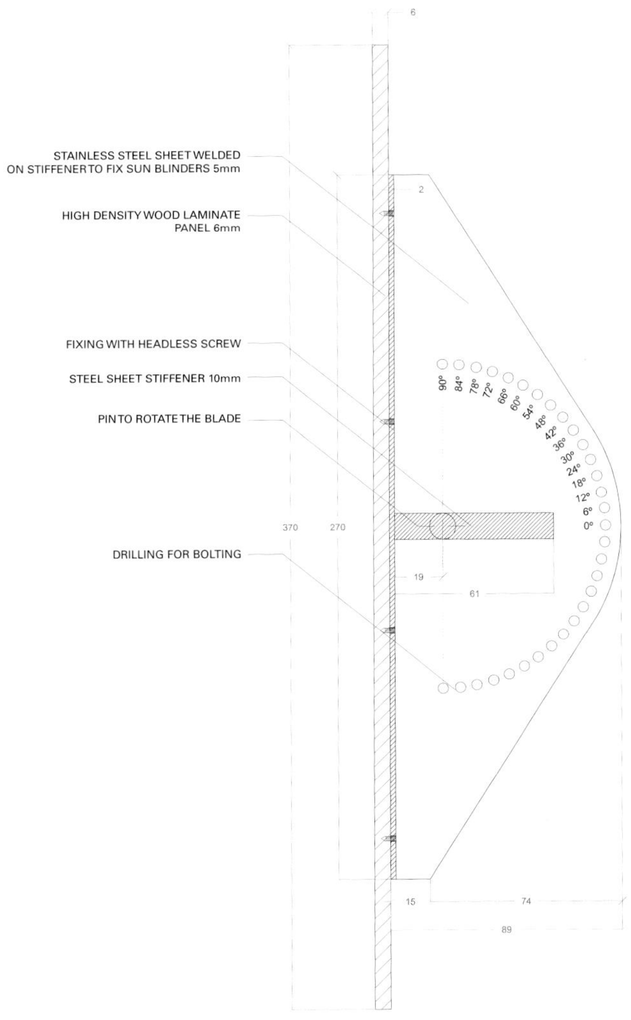

DETAIL HORIZONTAL SECTIONL 1/1°. Dimensions i

Detail of plate for shutters angle adjustment

WEAVING GLASS AND STEEL

Library, Auditorium, Exhibition Hall and Reserves Bayeux, 2020

Standing at the edge of the beltway encircling Bayeux's historic core, with its labyrinth of narrow, winding streets, the new Bayeux Media Library is carefully woven into the town's rich architectural and historical fabric. Located on an open site, the building acts as a link between the medieval center and the town's future expansion zones, offering unobstructed views of the cathedral. This exceptional setting demanded a sensitive architectural response. The design embraces transparency, creating a landscape building that serves as a living urban showcase. Open and inviting, the project fosters a fluid dialogue between interior and exterior, with a central patio and outdoor reading terraces on the south side, extending the library's spaces and activities into the open air.

Natural light is a key component of the library's design. Each space is illuminated through varied light sources, carefully orchestrated to create clarity and spatial legibility. Skylights punctuate the green roof, bringing daylight deep into the interiors, while the central patio channels light into the heart of the building, fostering a luminous and welcoming atmosphere. A perforated metal mesh on the north façade softens incoming light, ensuring gentle diffusion throughout the reading areas. This results in a highly transparent, open-access library where users intuitively navigate toward books and resources. The northeastern entrance, slightly set back from the urban intersection, welcomes visitors into a calm and inviting forecourt, away from the surrounding traffic.

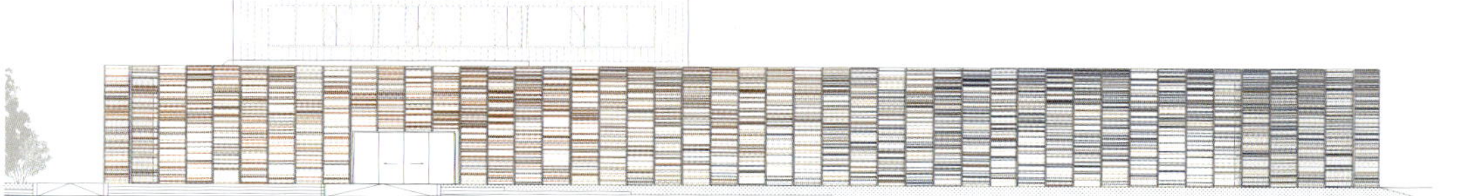

The history of Bayeux is inseparable from its most famous artifact, the Bayeux Tapestry—also known as the Tapisserie de la Reine Mathilde. Woven and embroidered in the 11th century, this masterpiece directly inspired the north façade of the media library. Just as the tapestry was created stitch by stitch, thread by thread, the library's north façade embodies this artisanal spirit. Facing Boulevard Ware, the entire glazed surface is shielded by a "filtering skin" of tinted vertical tubes. These elements reference the natural woolen hues of the tapestry—beige, brown, bronze green, blue-black, deep blue with yellow highlights—blurring the boundary between past and present. This embroidered façade filters views between the interior and the street and regulates light levels, enriching the ambiance of the reading areas.

The embroidered filtering skin works with the overhanging southern roof, ensuring visual comfort and controlled luminosity inside the library. This adaptive envelope plays an integral role in the building's thermal regulation. The façade opens through glazed bays on the east, south, and west sides, fitted with adjustable roller blinds, allowing users to modulate light and heat intake. The green roof envisioned as a fifth façade, is punctuated with skylights that enhance daylight penetration while enabling natural ventilation during mid-season months. This synergy between architecture, climate, and history transforms the Bayeux Media Library into a luminous and living cultural hub deeply rooted in heritage and contemporary innovation.

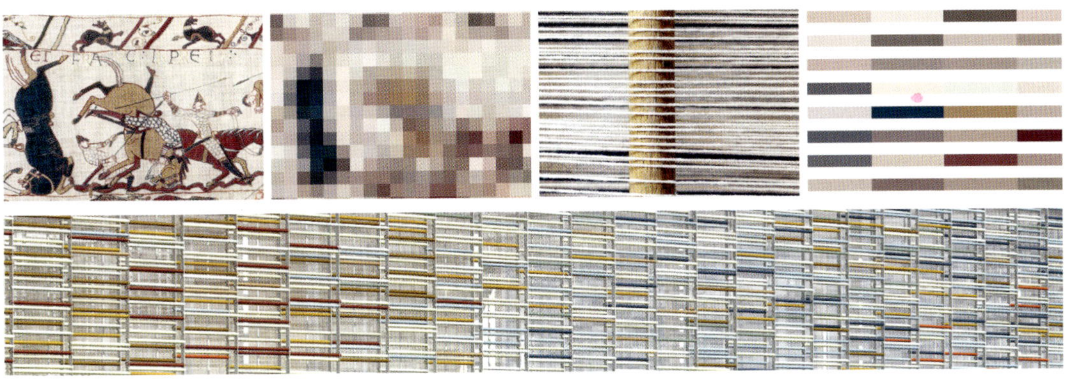

From Bayeux 12th century tapestry to woven color strands

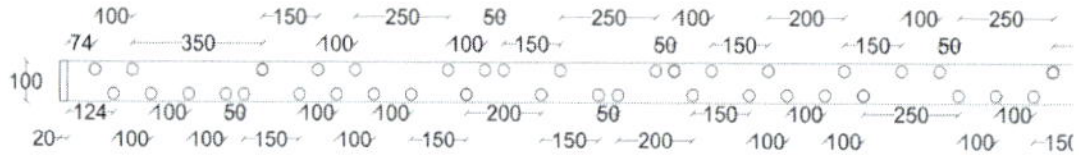

A1 type frame detail

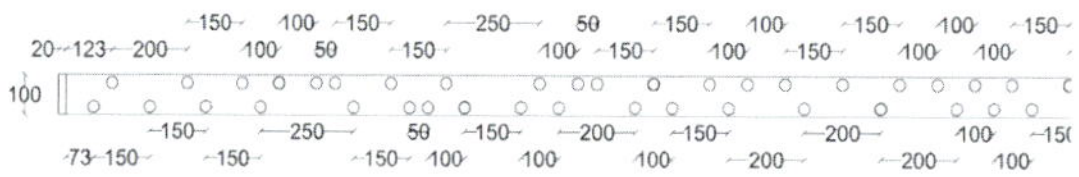

B1 type frame detail

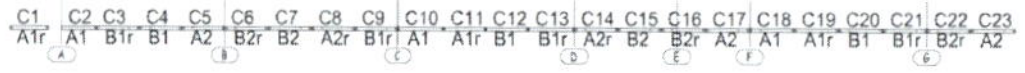

Frames distribution on north facade

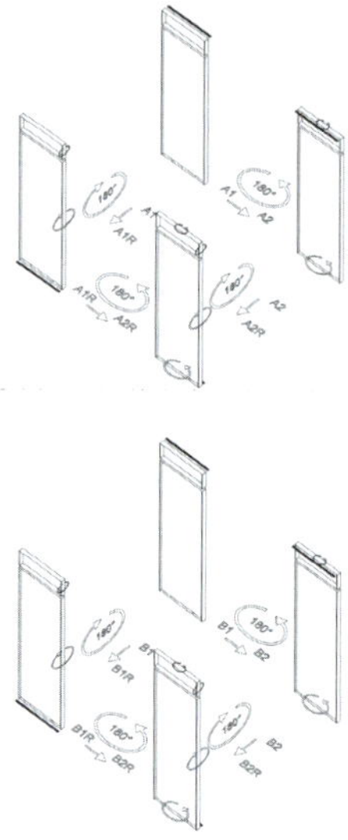

Declination diagrams of type A1 and B1 frames

Steel tubes facade woven screen

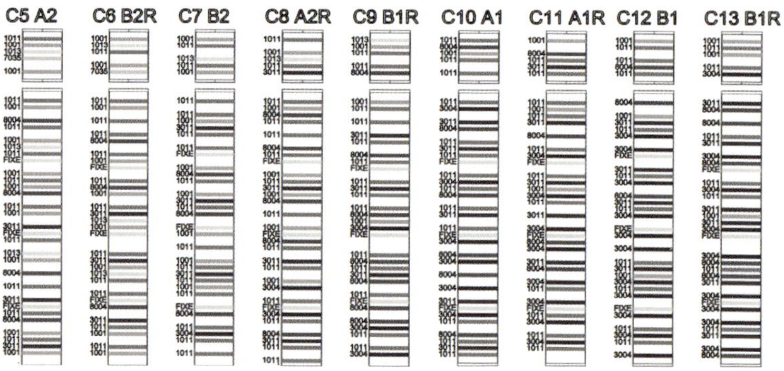

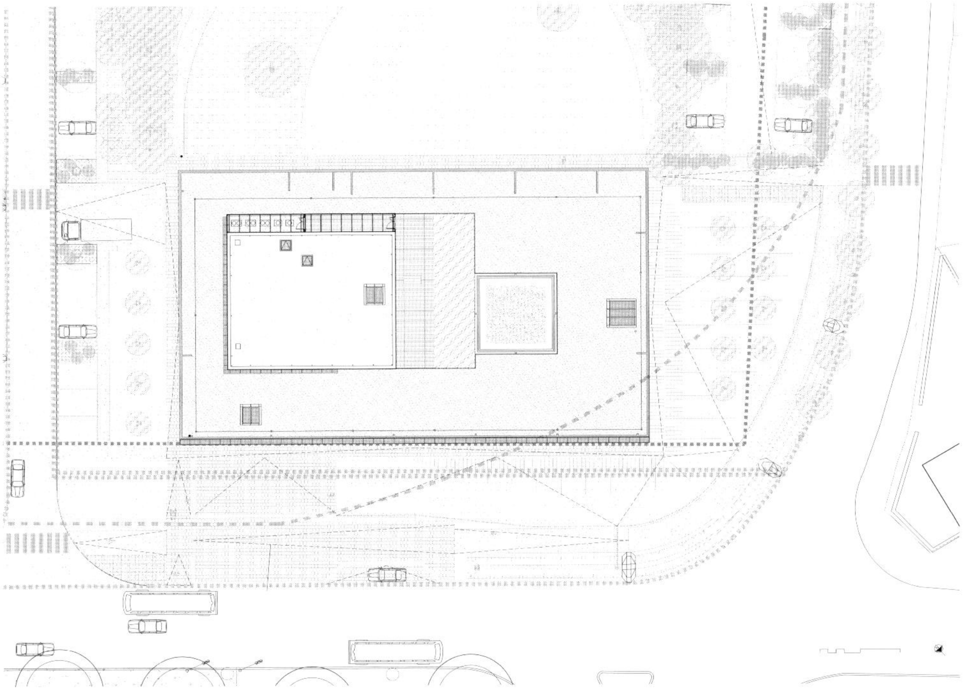

Building site plan

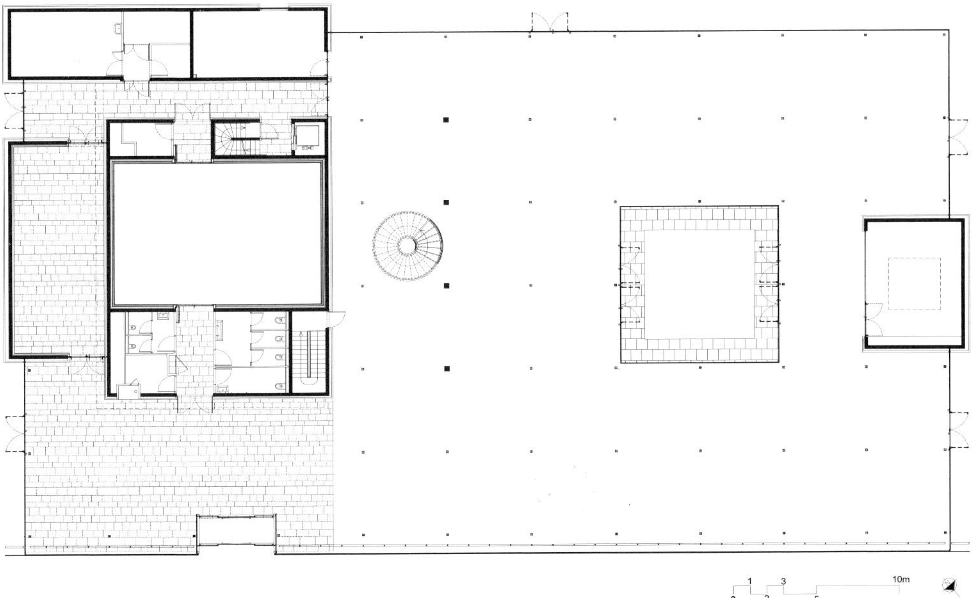

Ground floor plan layout

Night street view on Fabien Ware Boulevard, Bayeux

Garden facade on south side

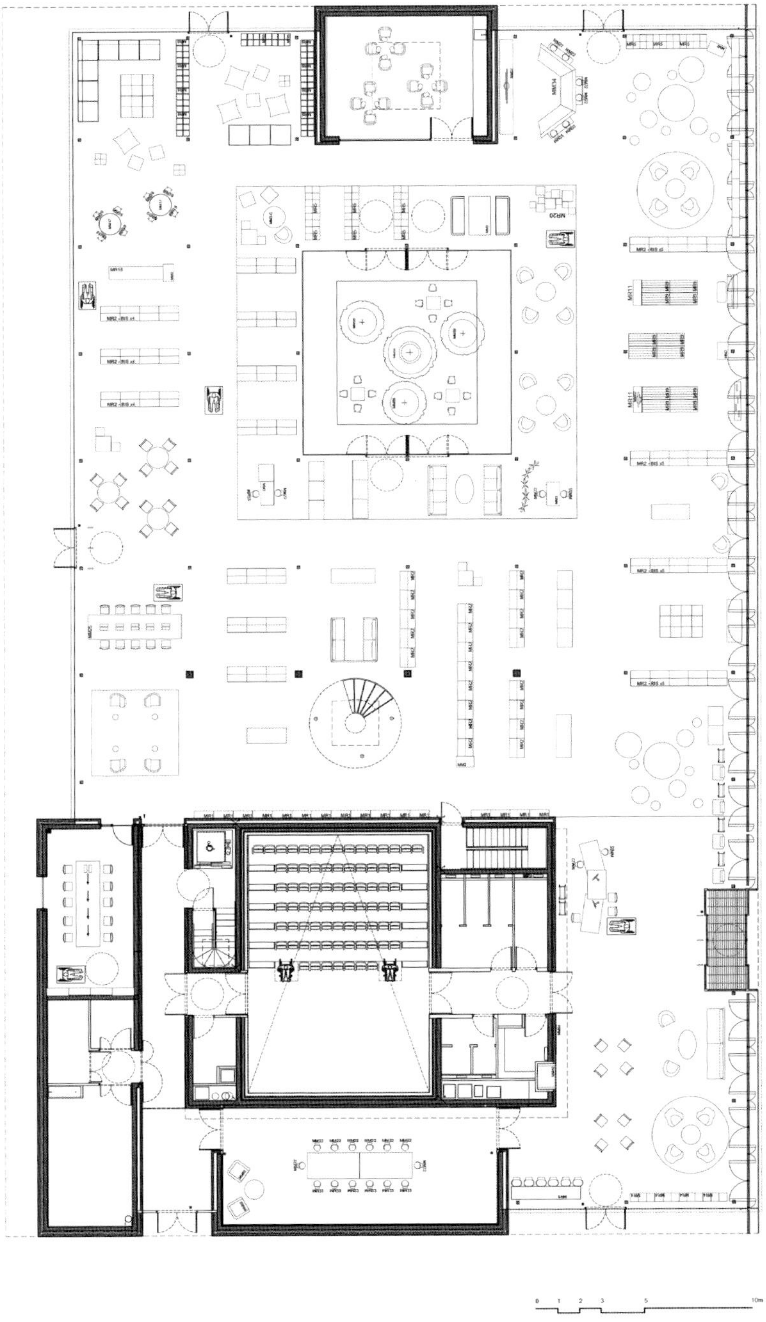

Ground floor plan

Comics reading saloon

docs jeunesse
Sciences
SCIENCES

CLIMAT PLANÈTE

Research Center, Laboratories and Data Center
Université Grenoble Alpes (UGA)- Grenoble, 2020

The Climat Planète Research Center is conceived as a twin building, extending the existing CERMO laboratory on the Université Grenoble Alpes (UGA) campus in Saint-Martin-d'Hères. Echoing the architectural DNA of its predecessor, the new structure is articulated around vertical circulations, with a rhythmic structural framework and a branching floor plan that organizes spaces into perpendicular volumes. More than just an extension, Climat Planète is designed to bring together a community of researchers, fostering exchange through its research infrastructure, Data Center, and digital computing virtualization room.

At the junction between the two buildings, "the Hub" is a fully glazed volume that reveals three levels of suspended platforms, housing reception areas, meeting rooms, and conference spaces. The hub serves as both a collaborative workspace and a showcase of research activity, offering visibility into the building's internal dynamics. To regulate light and maintain thermal comfort, the east and west façades are shielded by a woven stainless steel mesh, filtering sunlight while preserving transparency.

View of connected bridge between old and new campus

The office floors are open-plan and structured around a widened central circulation zone as a service corridor, integrating informal meeting areas, waiting zones, and natural ventilation systems. This layout fosters flexibility and adaptability, which are essential in an environment driven by innovation.

Designed to anticipate the rapid evolution of quantum and digital computing, the Data Center is conceived as a large open floor with a raised structure, allowing maximum reconfigurability. A supervision room and machine preparation space create a secure interface with the main platform, while a double-height technical room accommodates ventilation and cooling systems. Fresh air intake is managed through powder-coated aluminum louvers, ensuring energy-efficient thermal regulation.

The building's exterior is wrapped in a seamless envelope of aluminum panels, covering a bio-sourced thermal insulation system. A framework of structural frames accentuates the building's geometry, with angles and transparencies finely tuned to respond to each façade's orientation. Solar control is ensured by fixed vertical aluminum sunshades on the east and west façades, complemented by light shelves positioned above the windows to reflect natural daylight deep into the office spaces. Adjustable motorized sunshades further enhance the dynamic adaptation of interior lighting conditions.

Rooted in bioclimatic principles, Climat Planète integrates an intelligent energy management system, ensuring seamless interaction between the building and its environment. A natural ventilation strategy covers 90% of air renewal needs, minimizing energy consumption while leveraging the thermal inertia of concrete floors to regulate indoor temperatures. During summer and mid-season, airflow is driven by temperature differentials between the shaded ground-level air and the solar chimney on the roof, which enhances natural air extraction. In winter, a high-performance dual-flow ventilation system with energy recovery ensures thermal efficiency.

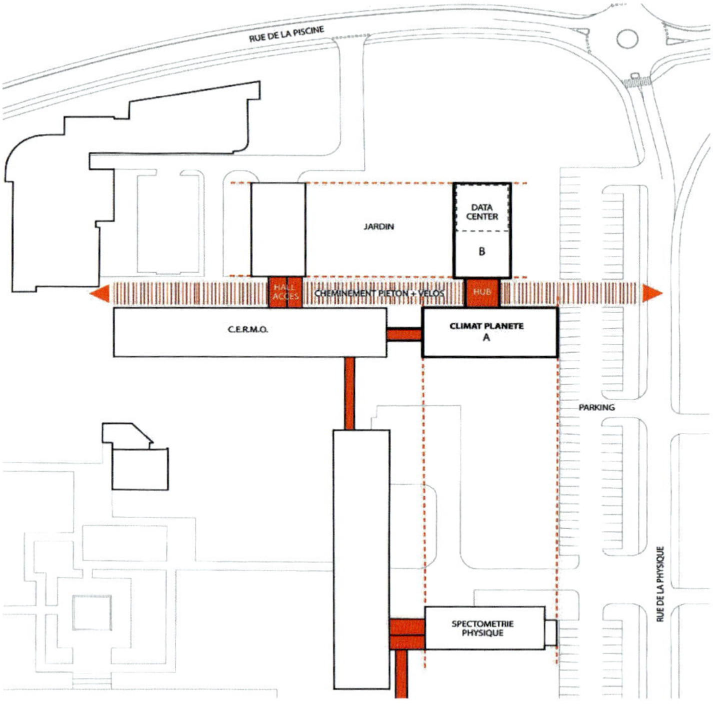

Campus circulation flow diagram

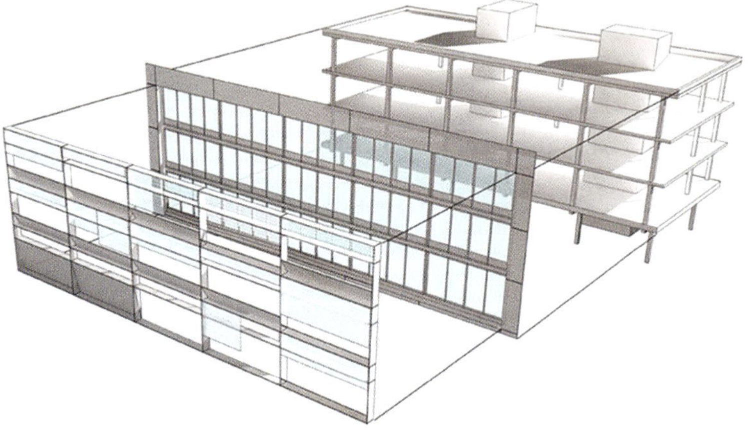

Structure to building enveloppe

Groundwater reserves allow for the implementation of geothermal wells, drawing water at 16°C to regulate indoor climate conditions. This system not only cools the servers and maintains comfort in the offices but also contributes to the hydraulic network of the underfloor heating/cooling system. Integrating a passive underfloor heating/cooling system that operates on low-energy water circulation further reduces energy demand.

The building features light shelves to optimize daylight use while controlling solar gain. These shelves diffuse natural light into workspaces while providing passive shading on the southern, eastern, and western façades. Server rooms and high-density computing spaces are cooled through a direct free-cooling system, which injects filtered outside air at the front of the servers, expelling excess heat efficiently to the exterior.
With its high-performance insulation, airtight construction, and advanced passive strategies, Climat Planète stands as the first positive-energy building on the UGA campus. It pushes the boundaries of sustainable architecture and redefines the research environment, creating a space where science, innovation, and environmental responsibility converge.

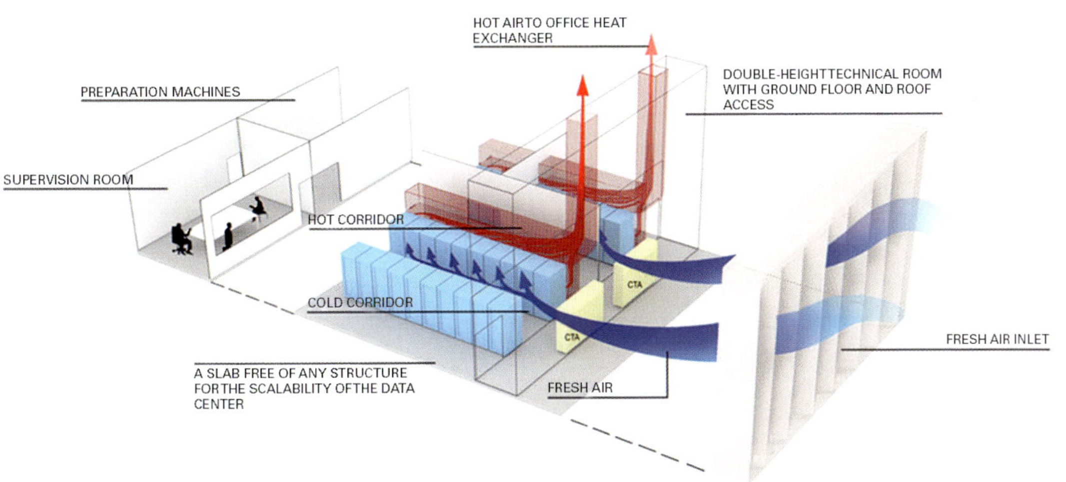

Datacenter geocooling system

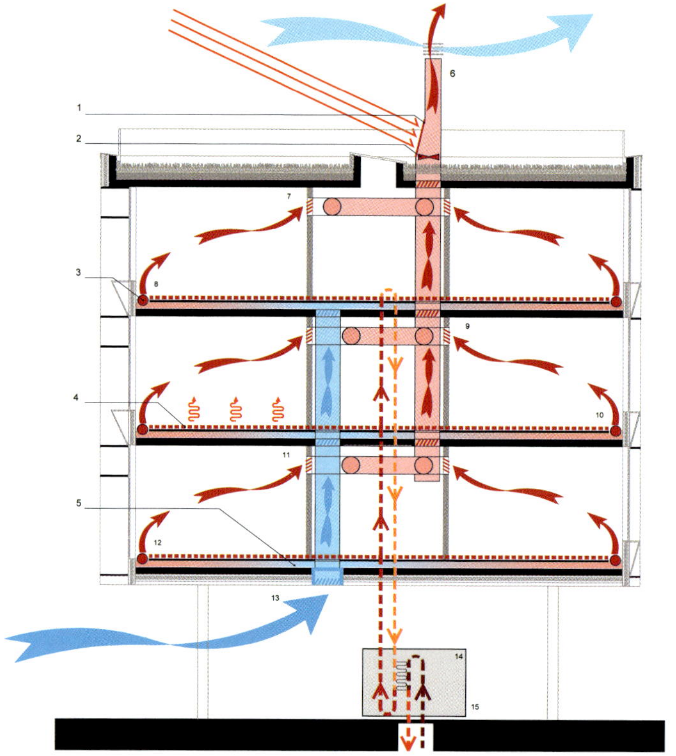

NATURAL VENTILATION WITH SOLAR ASSISTANCE

1 SOLAR PANEL SOUTH FACING
2 SOLAR FIREPLACE + FAN
3 HOT FINNED TUBE ON PAC CIRCUIT
4 FLOOR HEATING/REFRESHING ON PAC CIRCUIT
5 DOUBLE HOLLOW CORE SLAB FOR FRESH AIR DIFFUSION
6 HOT EXHAUST AIR
7 STALE AIR EVACUATION
8 AIR INTAKE NEW HEATED
9 STALE AIR EVACUATION
10 AIR INTAKE NEW HEATED
11 STALE AIR EVACUATION
12 AIR INTAKE NEW HEATED
13 FRESH AIR INTAKE
14 HEAT PUMP
15 GEOTHERMAL PUMPING ON WATER TABLE

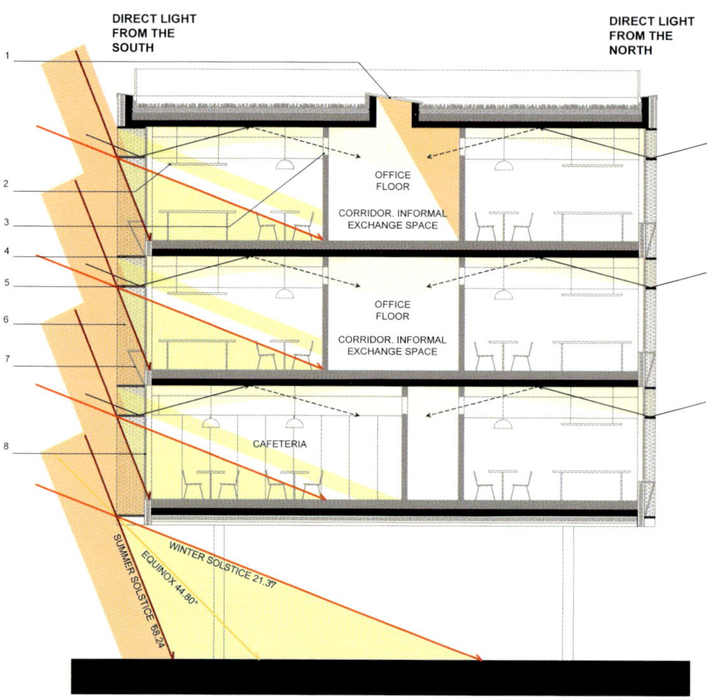

NATURAL LIGHT AND SOLAR PROTECTION

1 SKYLIGHTS FACING NORTH
2 LED HANGING LIGHTS CONTROLLED BY PHOTOSENSITIVE CELL
3 GLASS TRANSOM
4 PERFORATED ALUMINUM FIXED SUN SHADE
5 LIGHT SHELF
6 PERFORATED ALUMINUM FRAME (FRONT SYSTEM)
7 ALUMINUM CLADDING PERFORATED FOR LIGHT 100cm
8 SOUTH FULL HEIGHT WINDOW

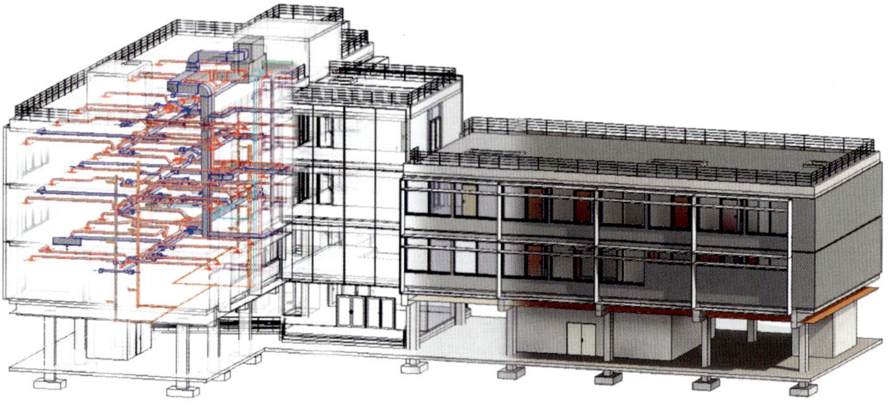

BIM integration of HVAC and technical systems

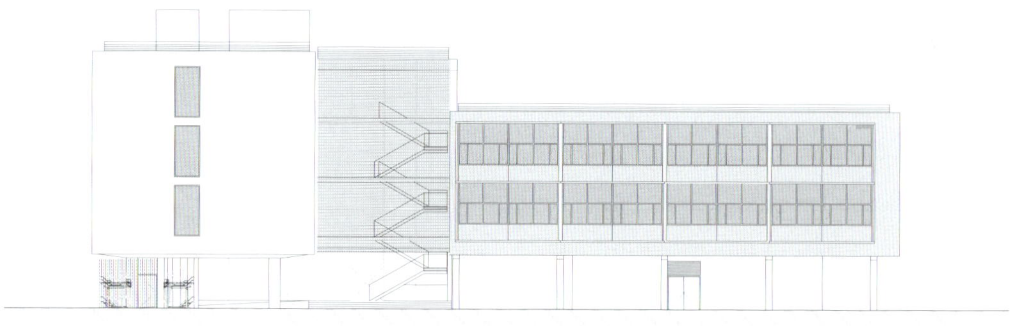

ELEVATION E-E' – EST

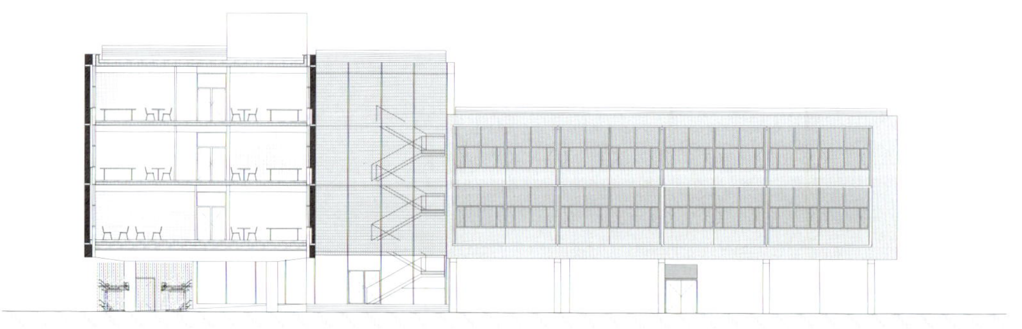

SECTION/ELEV F-F' – EST

Climat Planète building North and South elevations

Bio-climatic and interior environment treatment concept

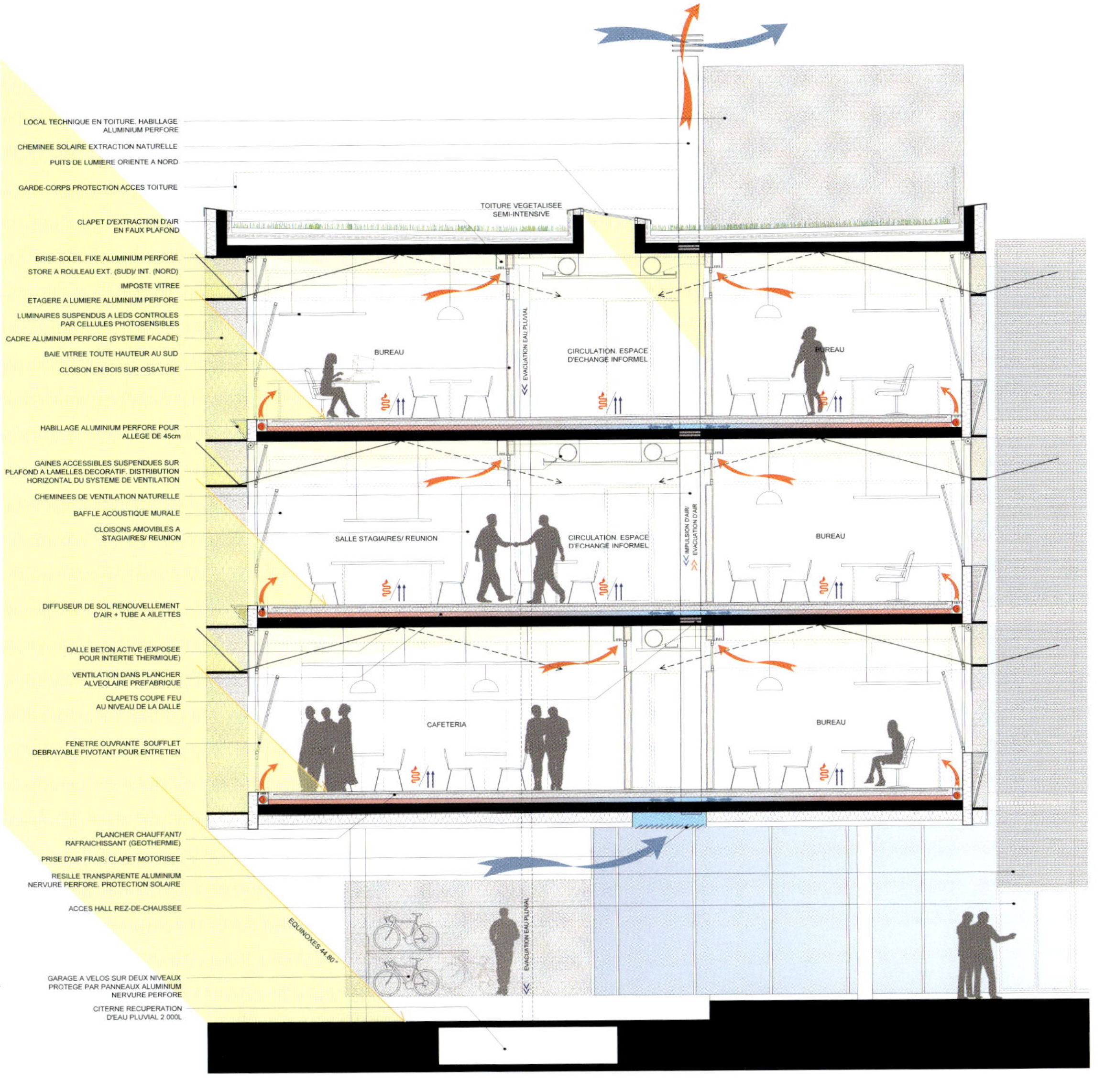

LOCAL TECHNIQUE EN TOITURE. HABILLAGE ALUMINIUM PERFORE
CHEMINEE SOLAIRE EXTRACTION NATURELLE
PUITS DE LUMIERE ORIENTE A NORD
GARDE-CORPS PROTECTION ACCES TOITURE
CLAPET D'EXTRACTION D'AIR EN FAUX PLAFOND
BRISE-SOLEIL FIXE ALUMINIUM PERFORE
STORE A ROULEAU EXT. (SUD)/ INT. (NORD)
IMPOSTE VITREE
ETAGERE A LUMIERE ALUMINIUM PERFORE
LUMINAIRES SUSPENDUS A LEDS CONTROLES PAR CELLULES PHOTOSENSIBLES
CADRE ALUMINIUM PERFORE (SYSTEME FACADE)
BAIE VITREE TOUTE HAUTEUR AU SUD
CLOISON EN BOIS SUR OSSATURE
HABILLAGE ALUMINIUM PERFORE POUR ALLEGE DE 45cm
GAINES ACCESSIBLES SUSPENDUES SUR PLAFOND A LAMELLES DECORATIF. DISTRIBUTION HORIZONTAL DU SYSTEME DE VENTILATION
CHEMINEES DE VENTILATION NATURELLE
BAFFLE ACOUSTIQUE MURALE
CLOISONS AMOVIBLES A STAGIAIRES/ REUNION
DIFFUSEUR DE SOL RENOUVELLEMENT D'AIR + TUBE A AILETTES
DALLE BETON ACTIVE (EXPOSEE POUR INTERTIE THERMIQUE)
VENTILATION DANS PLANCHER ALVEOLAIRE PREFABRIQUE
CLAPETS COUPE FEU AU NIVEAU DE LA DALLE
FENETRE OUVRANTE SOUFFLET DEBRAYABLE PIVOTANT POUR ENTRETIEN
PLANCHER CHAUFFANT/ RAFRAICHISSANT (GEOTHERMIE)
PRISE D'AIR FRAIS. CLAPET MOTORISEE
RESILLE TRANSPARENTE ALUMINIUM NERVURE PERFORE. PROTECTION SOLAIRE
ACCES HALL REZ-DE-CHAUSSEE
GARAGE A VELOS SUR DEUX NIVEAUX PROTEGE PAR PANNEAUX ALUMINIUM NERVURE PERFORE
CITERNE RECUPERATION D'EAU PLUVIAL 2.000L
TOITURE VEGETALISEE SEMI-INTENSIVE
BUREAU
CIRCULATION. ESPACE D'ECHANGE INFORMEL
BUREAU
EVACUATION EAU PLUVIAL
SALLE STAGIAIRES/ REUNION
CIRCULATION. ESPACE D'ECHANGE INFORMEL
BUREAU
IMPULSION D'AIR/ EVACUATION D'AIR
CAFETERIA
BUREAU
EQUINOXES 44.80°
EVACUATION EAU PLUVIAL

MAISON
CLIMAT
PLANETE

Ventilated double skin facade on the south and west sides

Stainless steel woven mesh for hub solar protection

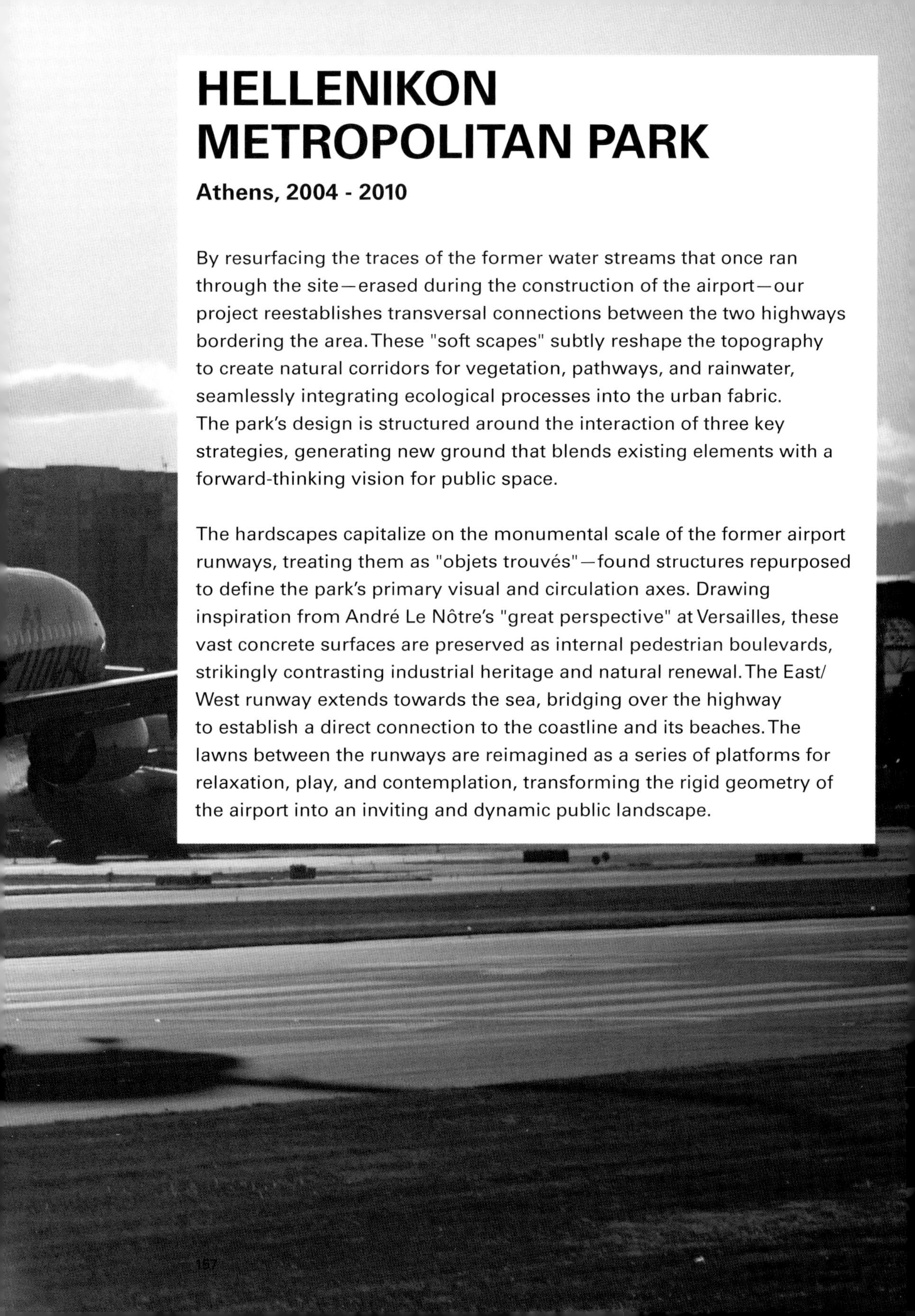

HELLENIKON METROPOLITAN PARK

Athens, 2004 - 2010

By resurfacing the traces of the former water streams that once ran through the site—erased during the construction of the airport—our project reestablishes transversal connections between the two highways bordering the area. These "soft scapes" subtly reshape the topography to create natural corridors for vegetation, pathways, and rainwater, seamlessly integrating ecological processes into the urban fabric. The park's design is structured around the interaction of three key strategies, generating new ground that blends existing elements with a forward-thinking vision for public space.

The hardscapes capitalize on the monumental scale of the former airport runways, treating them as "objets trouvés"—found structures repurposed to define the park's primary visual and circulation axes. Drawing inspiration from André Le Nôtre's "great perspective" at Versailles, these vast concrete surfaces are preserved as internal pedestrian boulevards, strikingly contrasting industrial heritage and natural renewal. The East/West runway extends towards the sea, bridging over the highway to establish a direct connection to the coastline and its beaches. The lawns between the runways are reimagined as a series of platforms for relaxation, play, and contemplation, transforming the rigid geometry of the airport into an inviting and dynamic public landscape.

The softscapes follow the natural irrigation patterns of the site, creating a network that links the city and the park through an integrated water management strategy. During heavy rainfall, water is channeled, delayed, and temporarily stored along these lines, utilizing a system of retaining walls, embankments, and terraces that continuously adapt to the surrounding topography. Rather than imposing an artificial landscape, this approach amplifies the site's existing ecological dynamics, fostering the emergence of a self-sustaining natural ecosystem that optimizes water distribution, enhances vegetation growth, and minimizes maintenance requirements.

The park edges maximize the interface between the city and the landscape, increasing interaction between urban life and green space while enhancing the real estate value of adjacent developments. Like the shifting boundaries of a river delta—where land and water intertwine in an endlessly evolving edge—the corrugated urban boundary extends the physical and visual relationship between the built environment and nature. A system of macro and micro undulations within this edge creates a dynamic inversion between city and park, blurring the line between urbanity and wilderness.

Aerial view of master plan principles - hard and soft scapes,
Hellenikon metropolitan, Athens

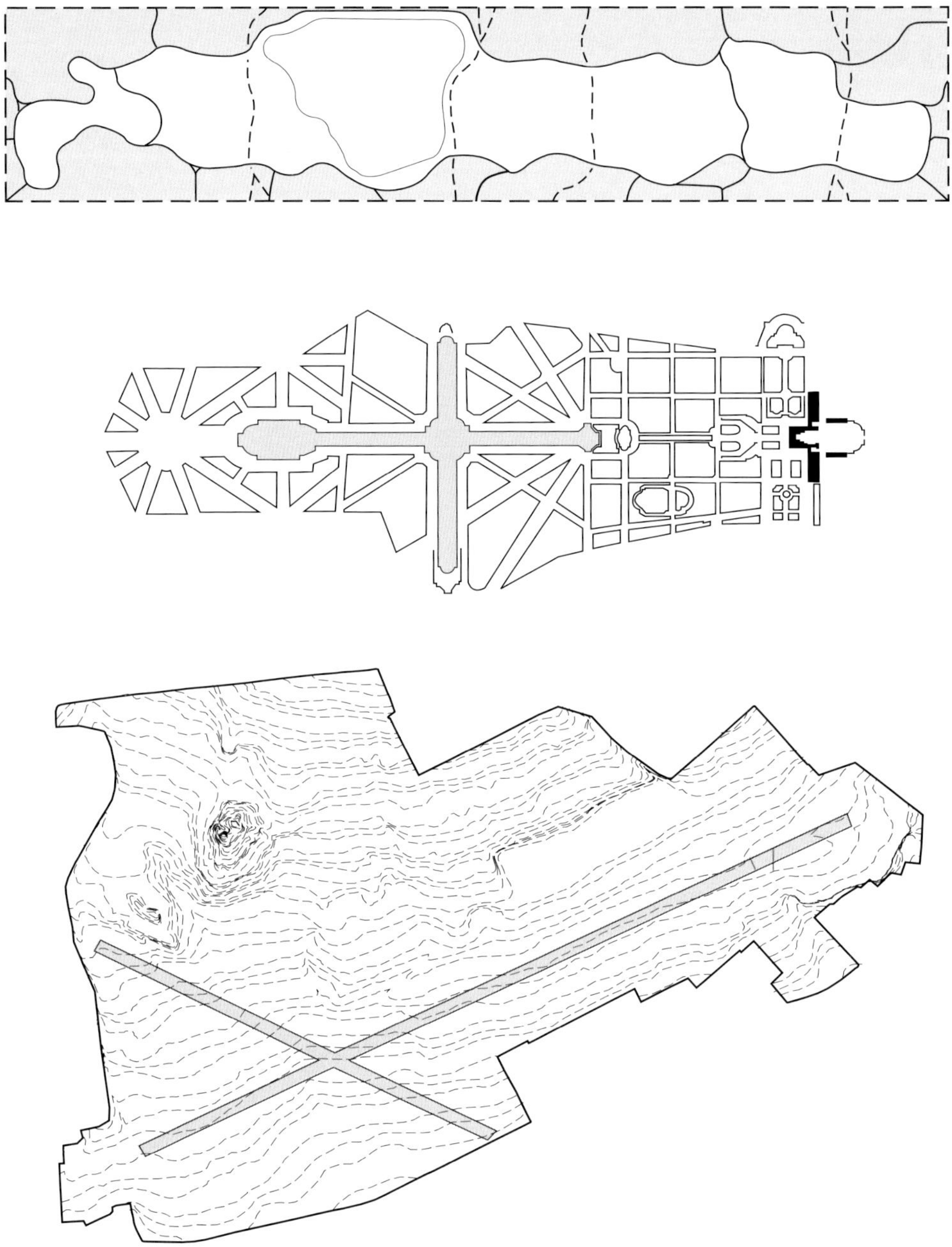

Scale comparison between Central Park - New York, Versailles Park - Paris and Hellenikon Metropolitan Park

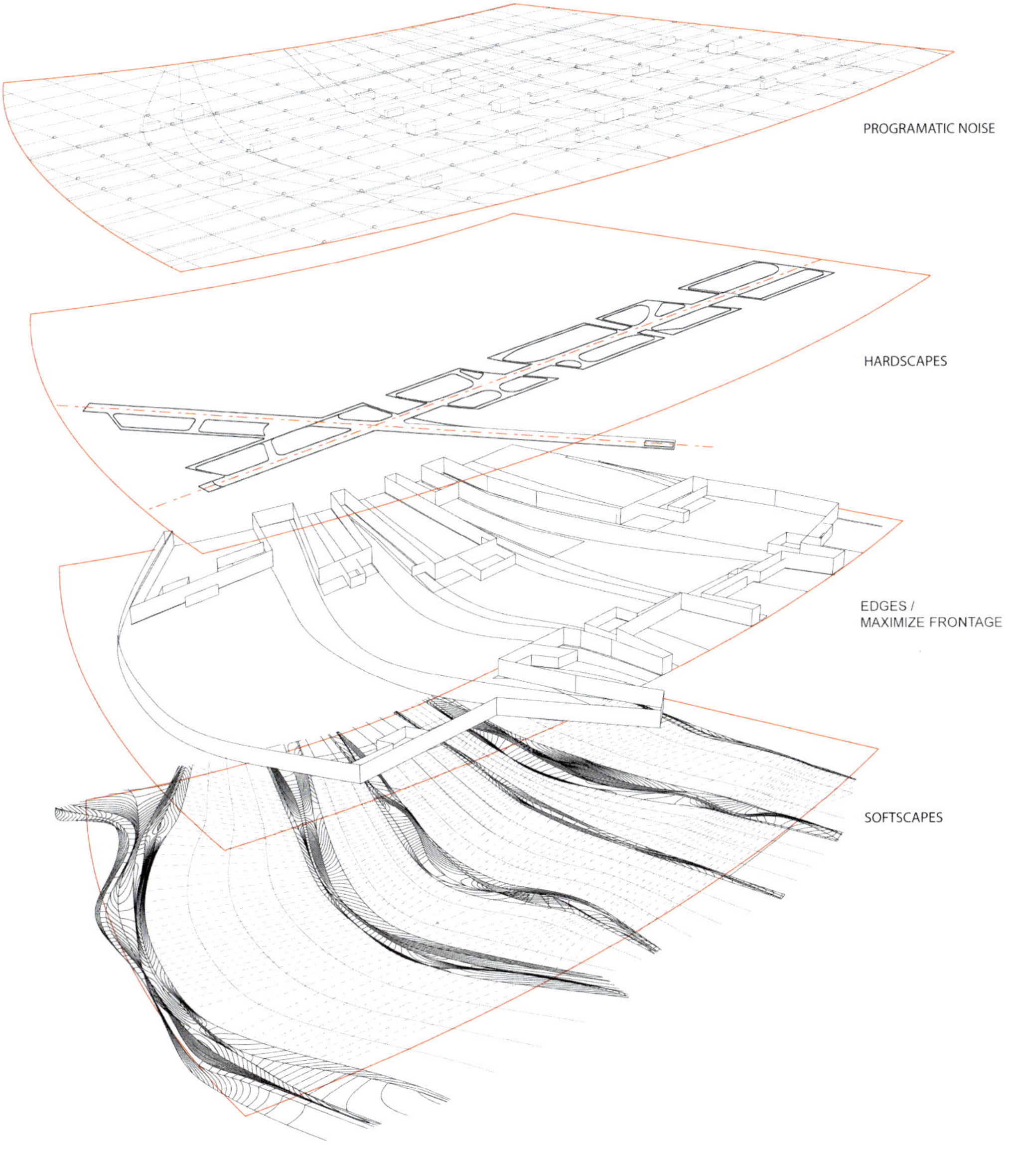

Conceptual axonometry of the park: Softscapes, Edges, Hardscapes, Programmatic "noise"

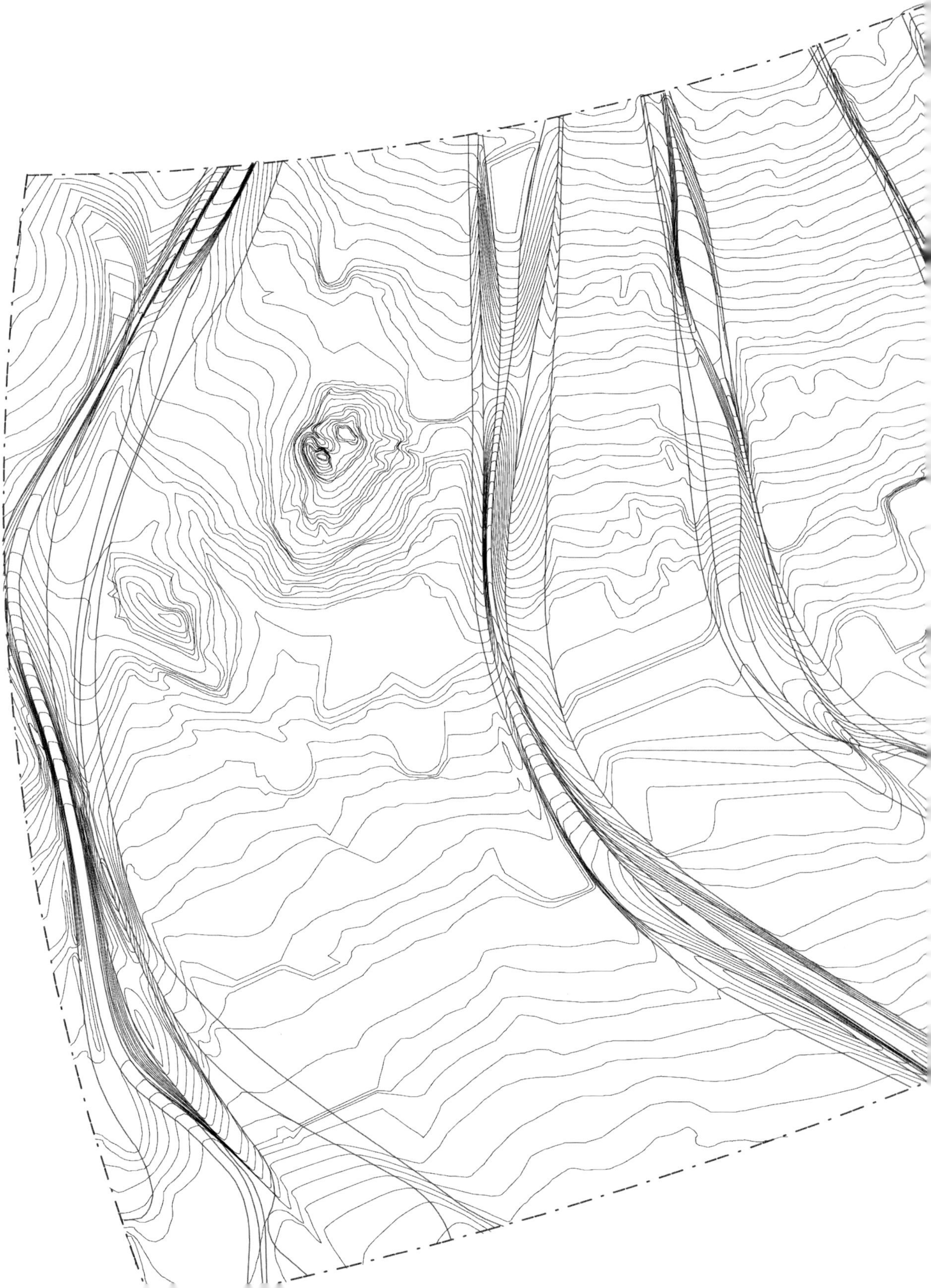

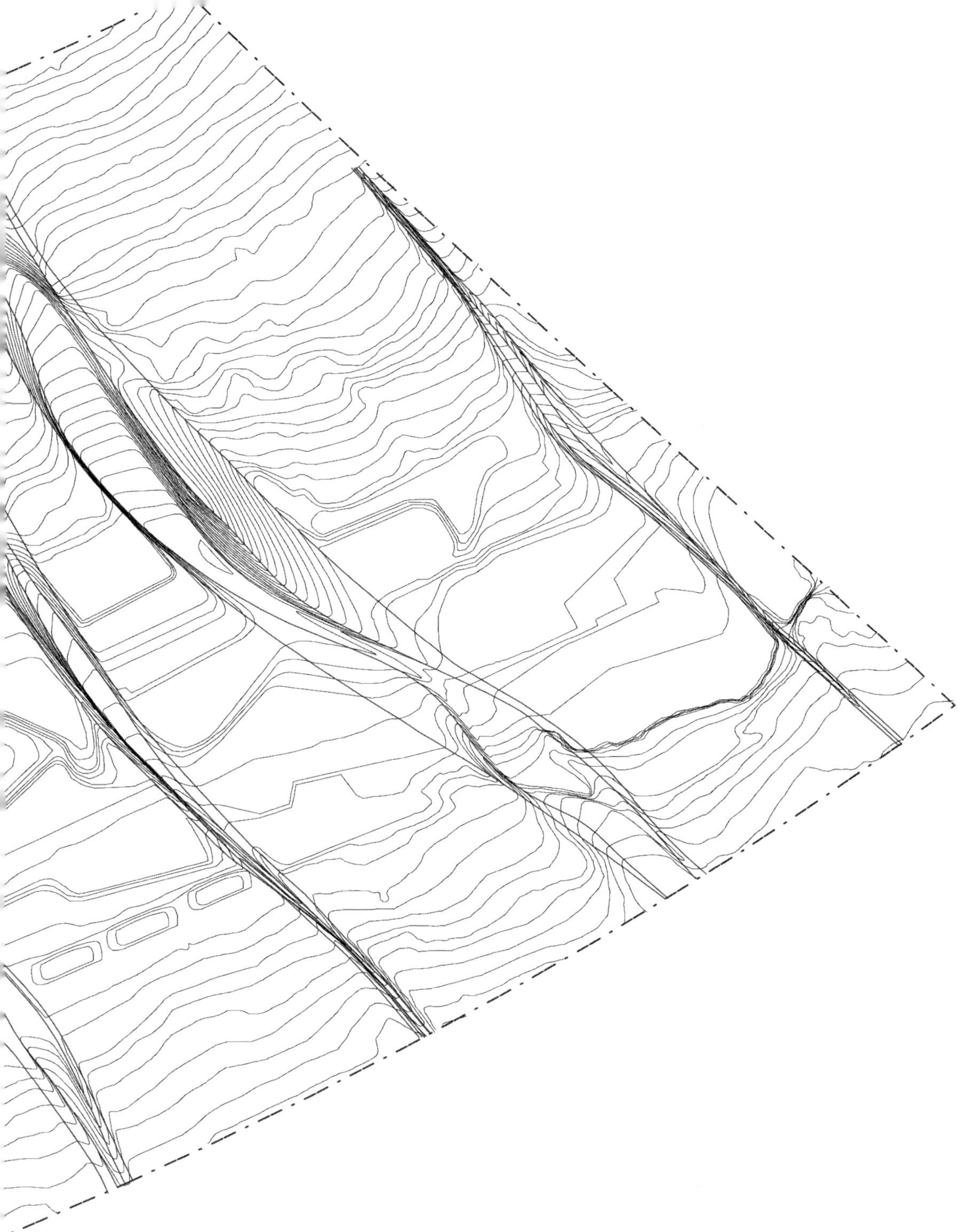

Topographic transformation along softscapes spines

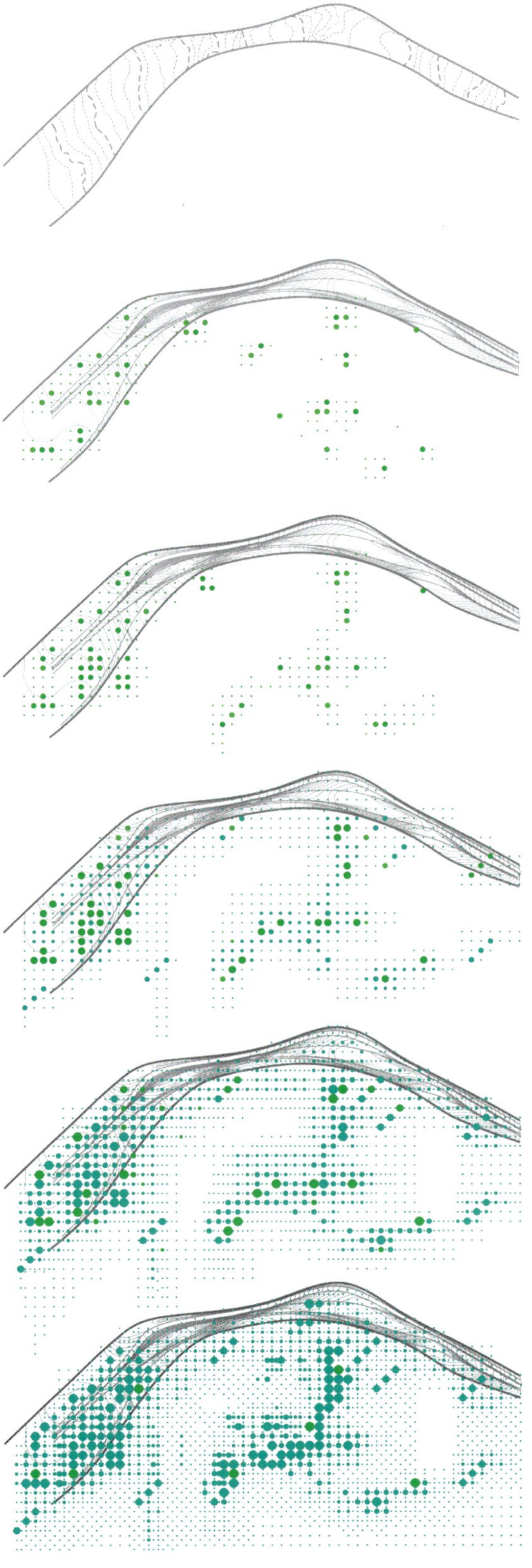

Self renaturalization of site
over a 40 years period

Led lighting matrice of Park at night, E. Redl

Landscape and programmatic layout for Hellenikon Metropolitan Park

POLE DANIEL ASSERAY

OPEN STACKS

Community and Social Center
Nantes, 2010

We took advantage of the unusual shape of the site on which the project is being built to create a unique building that is also a real focal point for the public spaces in the Dervallières district. Our project for the new administrative and social center occupies the maximum area of the site to develop a low-rise building close to the ground and easy to access, with a central reception area for the general public that is as large as possible.

In plan, its footprint is in the shape of an irregular pentagon, hollowed out in its center. The entrance to the building is in line with the main circulation axis of the district to facilitate its reading and access. The building with a large proportion of glass to offer the neighborhood a bright and welcoming space.

By creating a difference in levels to provide a horizontal reading of the building and an original screen print on the windows of the first floor, the building contributes to renewing the neighborhood's image. With the implementation of full or partial green roofs and the interior hanging garden, the project fits harmoniously into a site full of greenery. It is a "topographical" building, which functions using coherent programmatic entities that allow it to articulate the three functions of the building: Social services center, job search center, and local information center.

On the facade, we wanted each of these levels to be identifiable by having its facade system. The ground floor opens onto the city through large bay windows, the first level through a curtain wall on a parapet with a silkscreen print evoking the tree structures of the garden that we have suspended above the hall, and on the second and third floors, a system of cantilevers and slab overhangs to protect the windows from the sun and open the cafeteria onto a rooftop garden.

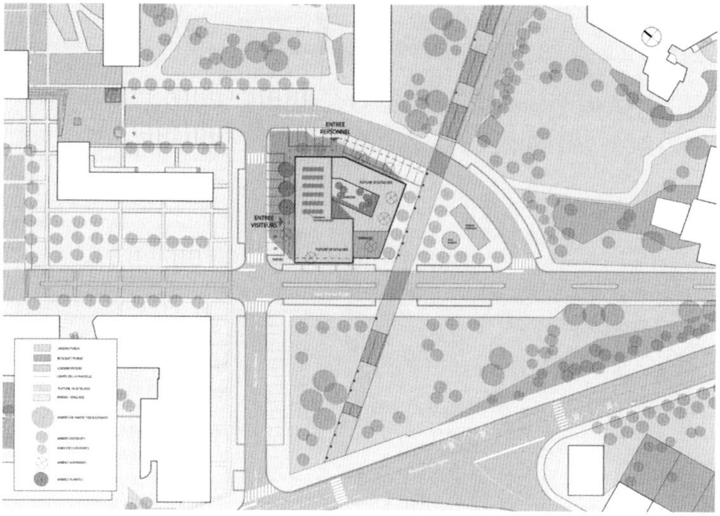

Site plan

Ground floor plan

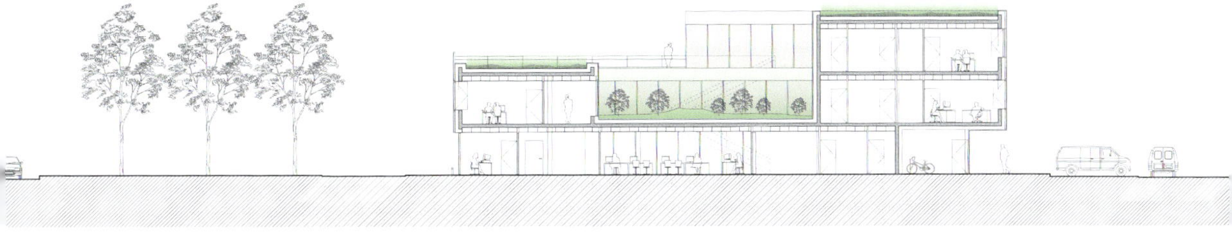

POLE DANIEL

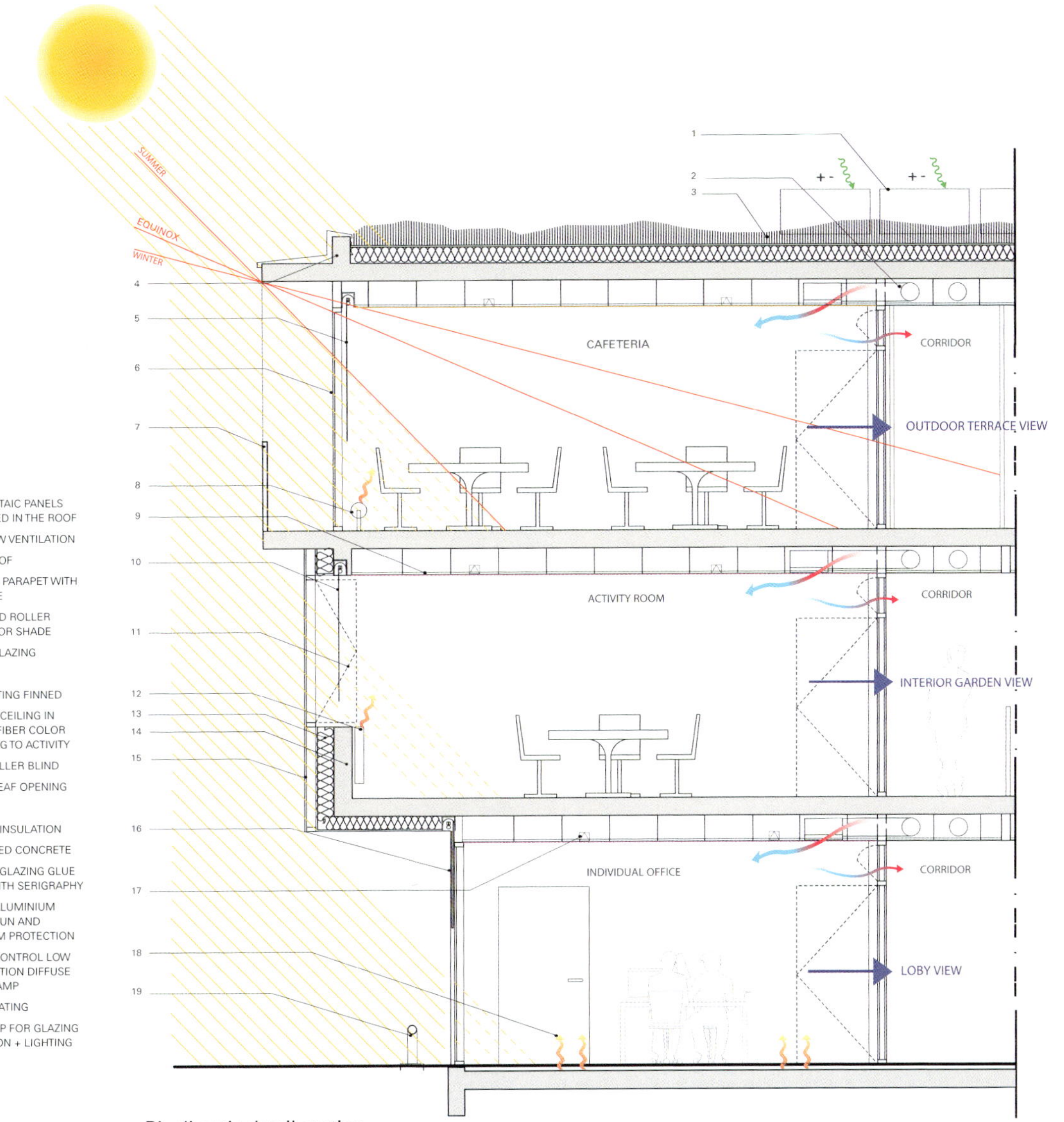

1 PHOTOVOLTAIC PANELS INTEGRATED IN THE ROOF
2 DUAL FLOW VENTILATION
3 GREEN ROOF
4 RECESSED PARAPET WITH SLAB EDGE
5 MOTORIZED ROLLER AWNING FOR SHADE
6 DOUBLE GLAZING
7 HANDRAIL
8 TUBE HEATING FINNED
9 ACOUSTIC CEILING IN NATURAL FIBER COLOR ACCORDING TO ACTIVITY
10 SHADE ROLLER BLIND
11 DOUBLE LEAF OPENING
12 HEATER
13 EXTERIOR INSULATION
14 REINFORCED CONCRETE
15 EXTERIOR GLAZING GLUE FACADE WITH SERIGRAPHY
16 OUTSIDE ALUMINIUM CURTAIN SUN AND VANDALISM PROTECTION
17 DIMMER CONTROL LOW CONSUMPTION DIFFUSE CEILING LAMP
18 FLOOR HEATING
19 STEEL STOP FOR GLAZING PROTECTION + LIGHTING

Bioclimatic detail section

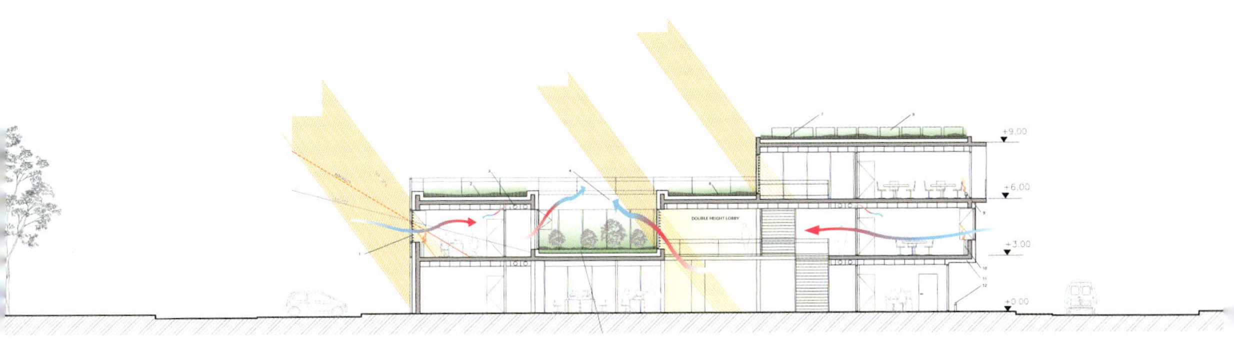

Stepped section of the building

INTEGRATED BIODIVERSITY IN CLUSTER AUDITORIUM

Restructuring Orléans University – Bourges, 2021

The University of Orléans campus in Bourges remains largely composed of buildings from the 1960s, which have seen little evolution since their construction. The challenge of this renovation was to transform these aging structures into efficient, contemporary learning spaces while significantly improving their energy performance. The project includes a building housing two amphitheaters, a 2,500 m^2 civil engineering department across three levels, a practical work hall, and a single-story learning center designed to foster innovation in pedagogical methods and sustainability.

One of the key focuses of the renovation was rethinking traditional lecture halls. How can a student remain engaged for several hours in a 300-seat amphitheater when studies show that attention typically declines after just 20 minutes? Considering this, the design introduced "cluster" amphitheaters structured to support lectures and group work. Each tiered seating row is divided into work islands for four to five students, featuring a curved bench and a shared digital tablet. This configuration enables students to collaborate in small groups while maintaining interaction with the lecturer. In close collaboration with faculty, this layout converts the amphitheater into a flexible, participatory learning space that accommodates interactive sessions and traditional lectures.

Campus entrance of auditoriums
and student services building

The renovation also integrated an innovative approach to biodiversity, recognizing the ecological potential of the campus. During the assessment of the existing buildings, a large bat population (approximately 120 individuals) was discovered, nesting within the polystyrene insulation cracks of the facades. Rather than displacing them, the project embraced a bio-inspired co-design strategy, ensuring the continued presence of these species while preserving the integrity of the building envelope.

The facade panels in composite stone cladding on external insulation incorporate custom-designed bat boxes that are seamlessly integrated into the architectural design. These long nesting slots are concealed behind the cladding and distributed strategically to ensure both accessibility and safety for the bats. To prevent any impact on users, a zinc shield was added below each nesting box to protect the facade and surrounding areas from droppings.

This ecological initiative was developed in collaboration with the Natural History Museum in Bourges, internationally recognized experts in bat conservation. Their input helped refine the nesting box dimensions, orientation, and microclimatic conditions, ensuring an optimal habitat while minimizing potential disruptions to the building's occupants. The design prioritizes stability, predator protection, and a discrete integration, making it a model for urban biodiversity preservation.

Since the building's inauguration, bat colonies have successfully recolonized the nesting boxes, with ecological surveys recording a significant increase in the local bat population. Within two years, the number of bats sheltering at the university has grown to around 200, compared to an estimated 110 before the renovation. Despite their presence, the exterior insulation remains intact, demonstrating the effectiveness of the design strategy in balancing wildlife conservation and architectural resilience.

This project represents a pioneering example of how architectural renovation can harmonize pedagogy, sustainability, and biodiversity, creating an innovative and ecologically responsible campus.

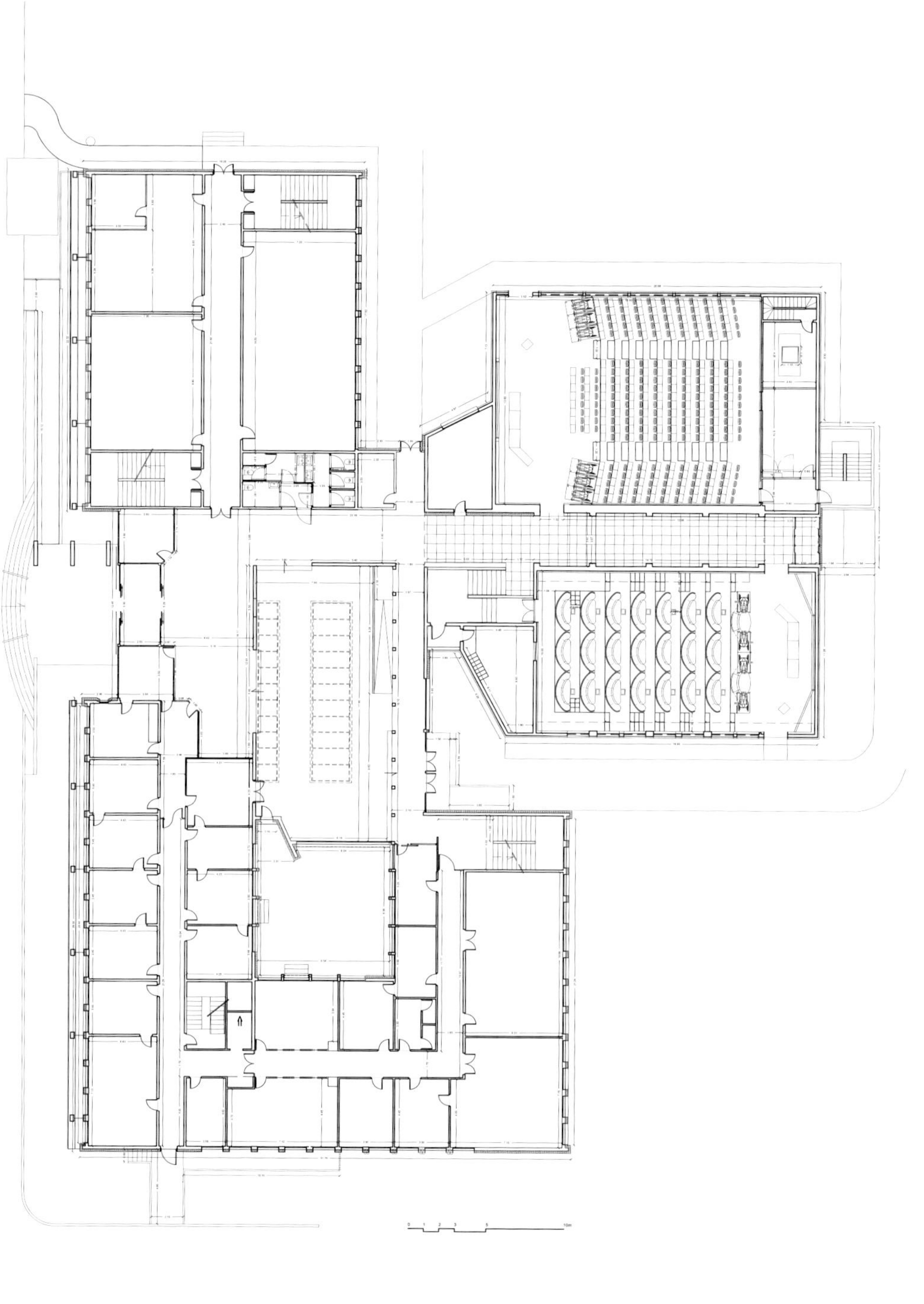

Ground floor plan

New entrance of experimental research hall

Cluster auditorium for experimental pedagogies

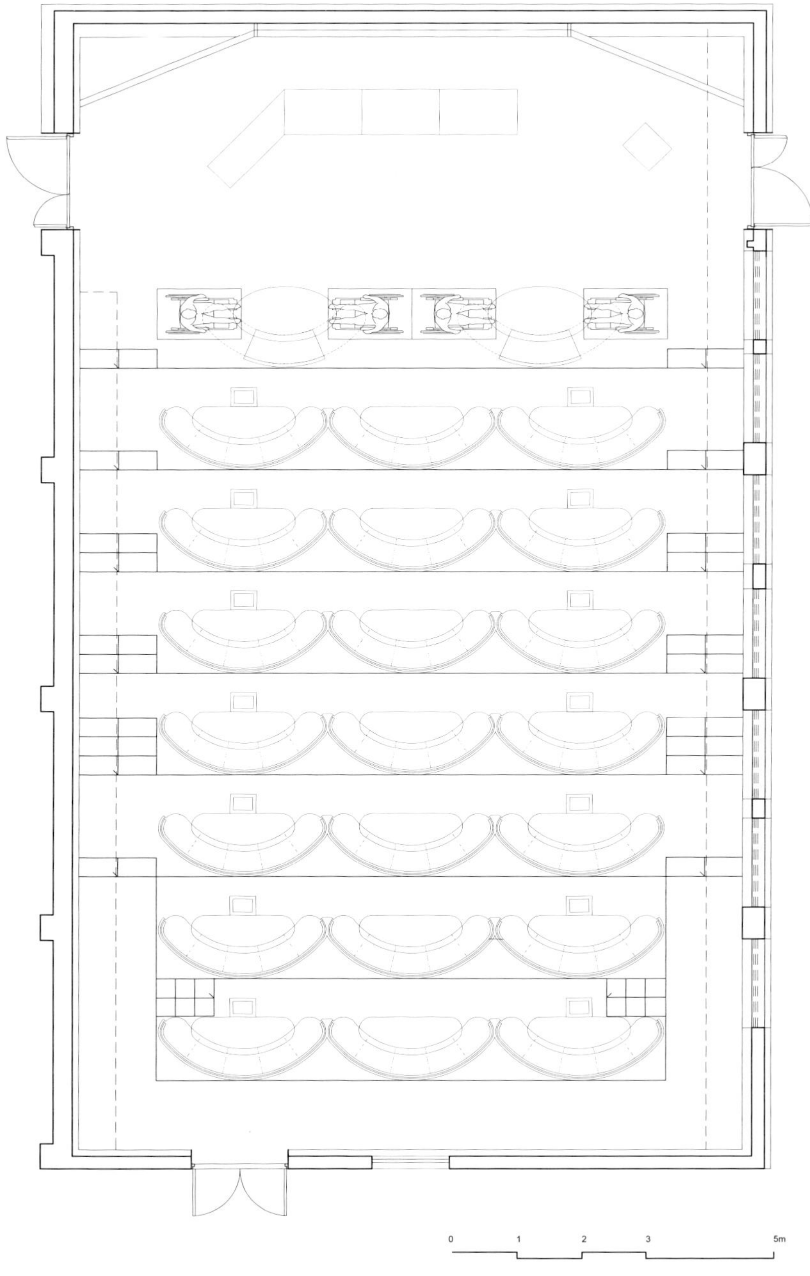

Cluster auditorium plan with 5 students
cells for collaborative work

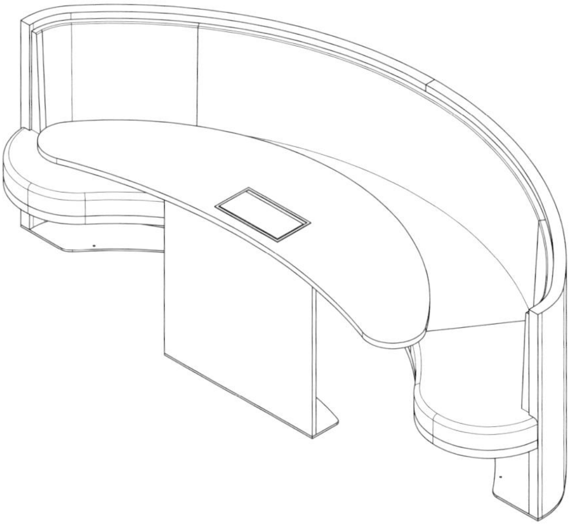

Detail of cluster

Main "wood" auditorium

Cluster auditorium

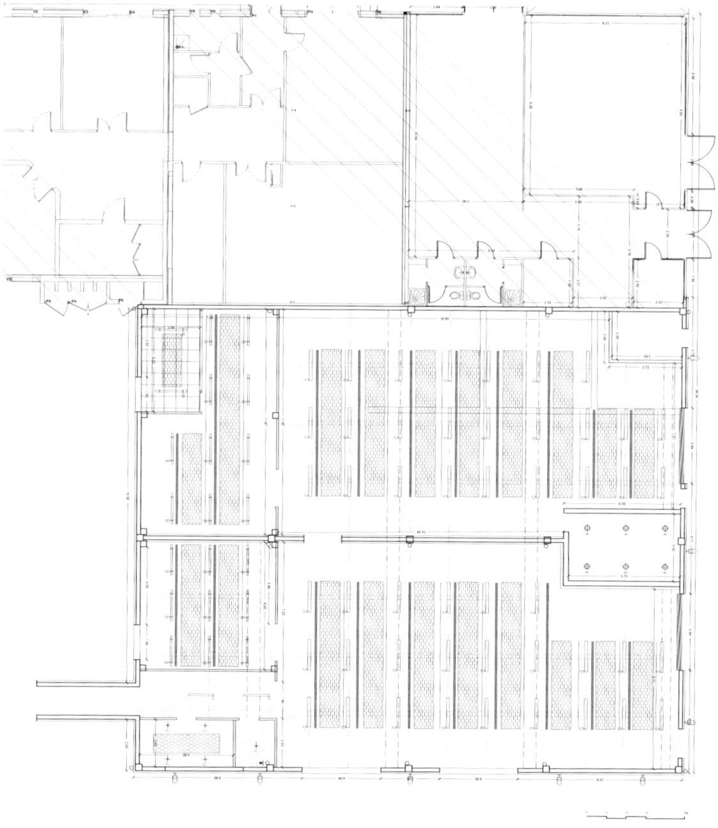

Experimental construction hall plan

Engineering school's experimental construction hall

Simulation Models for Energy-Less Architecture

We are fully committed to encouraging architects and designers to integrate energy-harvesting devices to balance building energy consumption. This approach aims to minimize maintenance costs and meet net-zero energy criteria. We often wonder how architectural design methodologies would evolve if a precise global energy efficiency study could map the wind patterns around a projected building. In response to this, could urban regulations be adapted to optimize airflow by modifying building profiles?

The research conducted by our team proposes a clear and effective design methodology for integrating wind-harvesting energy devices into architectural projects. By analyzing airflow around different architectural typologies, we explore the impact of urban environments on wind behavior. The study aims to identify optimal locations for wind turbines by leveraging an accelerating wing—referred to as an "aerofoil"—to enhance energy extraction. This methodology considers building orientation, scale, contextual conditions, and on-site wind data to optimize building-integrated device (BID) performance, ensuring efficient resource utilization across building envelopes.

Contextual Wind Data Analysis

Wind direction and velocity significantly affect extractable wind power. In urban environments, buildings and obstacles create complex airflow patterns. Factors such as building shape, height, and spacing influence wind flow direction and intensity, thereby impacting potential energy production.

Airflow within urban canyons is governed by street geometry and building height distribution. Due to aerodynamic drag, wind speed is typically higher when flowing parallel to a river but decreases when perpendicular. Computational Fluid Dynamics (CFD) simulations were combined with on-site wind measurements to establish an accurate urban wind cartography. This method revealed the profound impact of urban volumes and obstacles on wind speed and direction, highlighting the difficulty of predicting wind patterns without precise modeling.

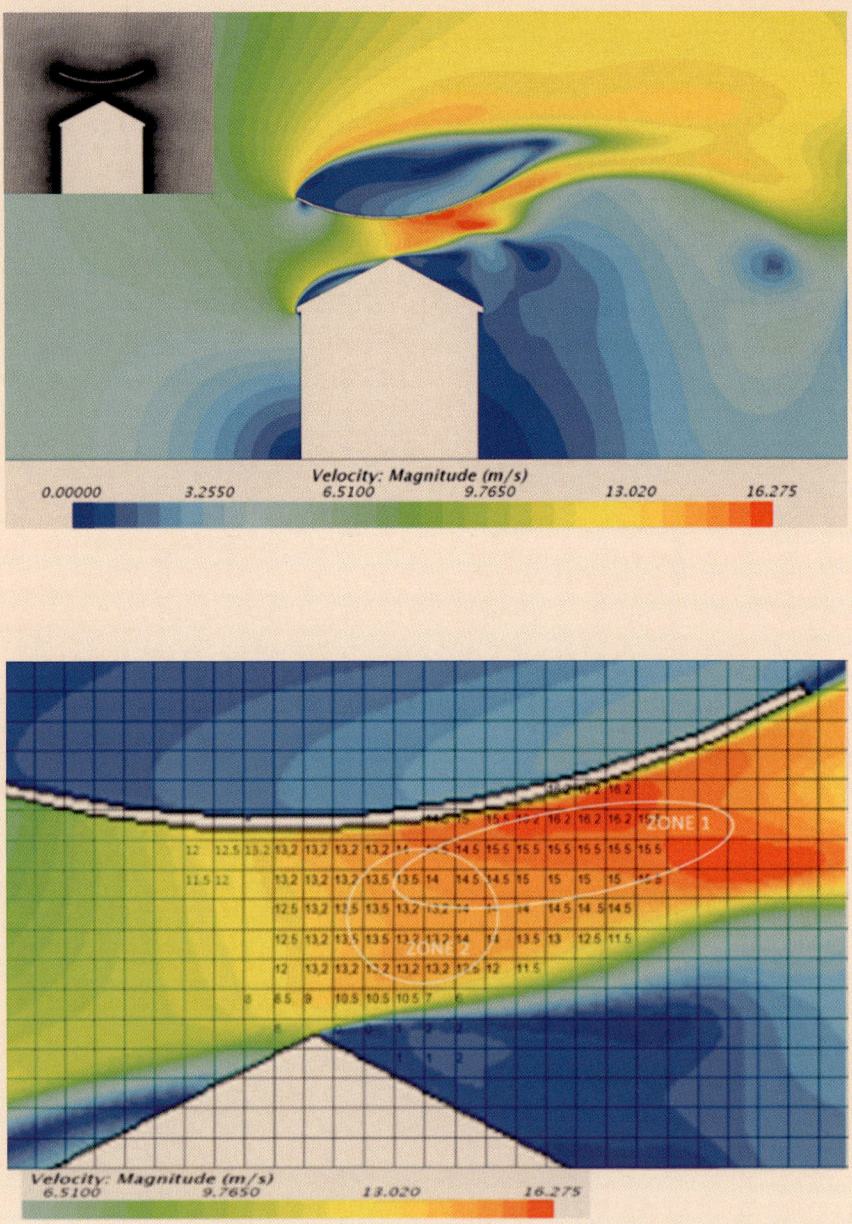

Aeraulic simulation of the airfoil located above a house roof

Instead of relying solely on theoretical wind analyses, this study incorporates data from connected anemometers to establish localized maps of urban winds. These surveys quickly demonstrated the strong impact of urban morphology on airflow, showing how wind speed and direction can increase or decrease significantly around buildings. The variability in wind behavior, influenced by urban blocks and obstacles, makes it challenging to predict patterns without precise simulations. The study highlights how urban configurations can create zones of high turbulence and multidirectional wind forces within a compact urban footprint.

Simulation Analysis of an Urban Digital Twin Model

We developed a volumetric 3D model of an urban block through extensive architectural practice to evaluate wind behavior on clustered rooftops in dense areas. This method allows for:

–Categorization of different wind types in urban settings
–Identification of minimum and maximum wind speeds
–Determination of optimal locations for energy-harvesting devices

We measured wind speed enhancement by integrating on-site wind analysis with CFD simulations, which directly influences extractable energy (proportional to the cube of wind speed, according to German physicist A. Betz). The simulation virtual environment allows for extensive parameter customization, ensuring accurate simulations that align with real-world measurements.

Many years of experience in architectural simulation have allowed us to refine our approach, creating models that accurately depict urban wind behavior. The volumetric model accounts for altimetry variations, spatial orientation, and building density, all of which influence wind acceleration and turbulence. The goal is to enhance energy harvesting by determining the best locations for wind turbines within an urban context. Additionally, the research investigates how wind acceleration phenomena occur at specific breakpoints in the urban fabric, such as roof edges and building corners, where wind speeds can be significantly amplified.

Aerofoils and Architectural Integration Typologies

"Aerofoils," inspired by nautical engineering, regulate and accelerate wind speed at building edges, leveraging the Bernoulli principle and Venturi effect. Our study identified key architectural elements—such as

	MICROTYPE	STRAIGHT	HELLICOIDAL	CLUSTER/ AGGREGATION
DARRIEUS			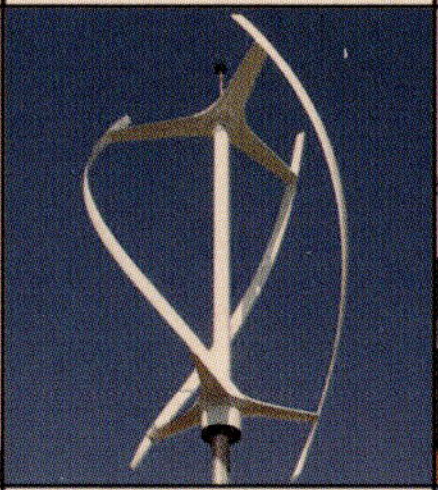	
WITH BUCKETS				
WINDSPIRE				
SAVONNIUS		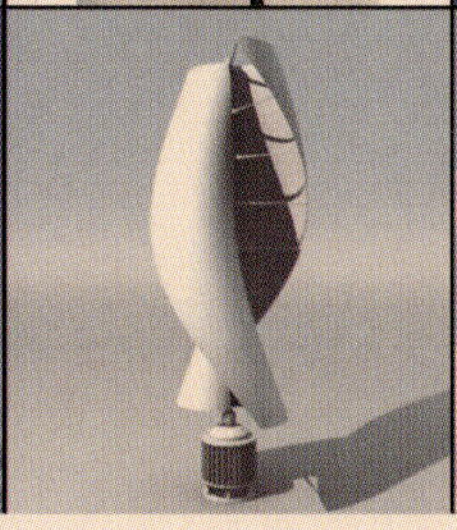		

Typologies of micro-wind. The vertical axis is privileged to adapt to the turbulent wind regime in urban sites

canopies, awnings, and interstitial spaces—that naturally concentrate airflow, making them ideal for energy harvesting.

Micro-wind turbines, classified by their swept area ($<25m^2$) and rated up to 6 kW, can significantly benefit from urban wind amplification. Under our simulation conditions, wind power density increased by a factor of 3 to 8, while wind speed amplified by 1.5 to 2 times due to building-induced effects.

The study further analyzed how integrating wind-accelerating devices like aerofoils can improve airflow stability around buildings. These elements mitigate turbulence and enhance wind turbine efficiency by concentrating and directing wind currents. Aerofoils act as control surfaces that modulate airflow, allowing architects to design structures that actively contribute to energy generation. Additionally, we explored how different BID typologies influence energy production, considering variations in height, wind exposure, and urban density.

The study further analyzed how integrating wind-accelerating devices, such as aerofoils, can improve the stability of airflows around buildings. By concentrating and directing wind currents, these elements mitigate turbulence and enhance wind turbine efficiency. Aerofoils act as control surfaces that modulate airflow, allowing architects to design structures that actively contribute to energy generation. Additionally, we explored how different BID typologies influence energy production, considering variations in height, wind exposure, and urban density.

Aerofoils, originally developed for maritime and aeronautical applications, have proven effective in architectural contexts. By examining how airflow behaves along building surfaces, we identified zones where energy-harvesting devices can be most effectively placed. The study also discusses the integration of aerofoils with façade-mounted turbines, demonstrating how these elements can stabilize turbulent airflow and increase wind capture efficiency.

Integration of Energy Harvesting Devices in Architectural Design

Urban turbulence necessitates strategic turbine placement to ensure stable power output. Integrating BID devices into architectural forms can be categorized as Standalone wind turbines (with simplified installation and maintenance)or retrofitted turbines (with Performance

Wind anemometer installation in a Paris building

Water and airfoil typologies

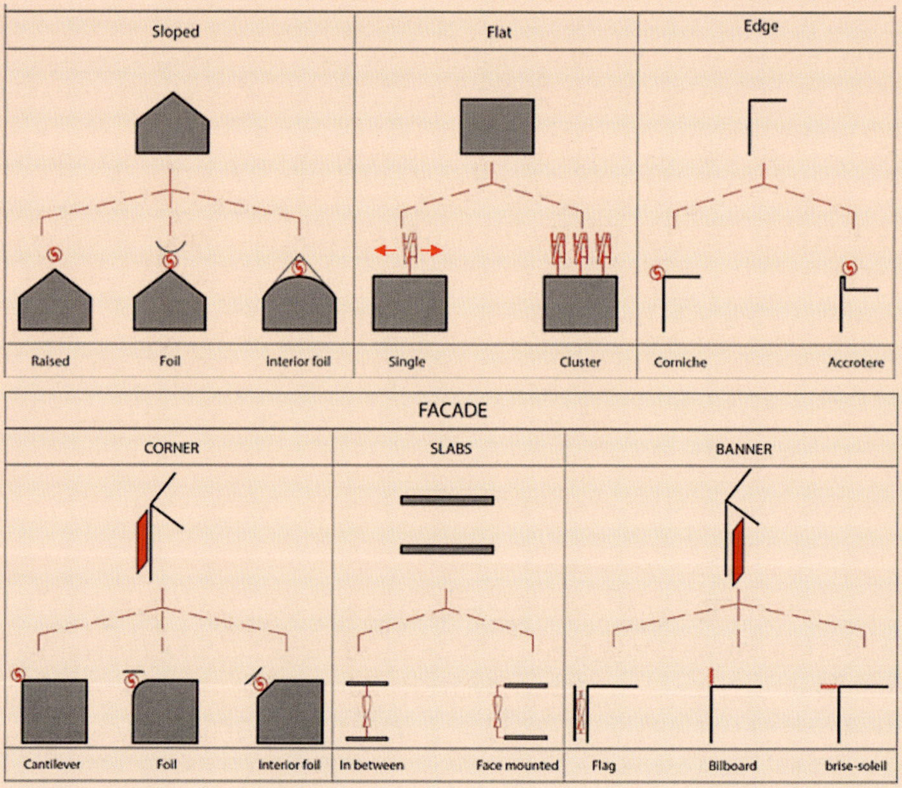

Wind micro-turbine installation typology

varying based on positioning, structural stability, and vibration dissipation).

To validate our methodology, we analyzed the ENSA Paris-Val de Seine architecture school, designed by Frédéric Borel in 2008. This building was selected for its block-like structure, strategic location in a high-wind area, and accessibility for wind speed monitoring. CFD simulations showed significant wind amplification between its two towers, increasing speeds from 5 m/s to approximately 17 m/s.

To harness this effect, we proposed rigid aerofoil panels to concentrate and stabilize wind flow. By adjusting aerofoil angles, wind speeds between the towers were optimized to 16–18 m/s, leading to an estimated compensation of 18% of the building's annual energy consumption through BID integration.

Furthermore, the study explores potential strategies for optimizing urban wind energy production by considering new approaches to façade design. Rather than treating building surfaces as passive elements, this research advocates for active integration, transforming urban architecture into an energy-generating ecosystem. Using wind amplification techniques combined with BID elements allows for more efficient resource utilization, contributing to long-term sustainability in high-density environments.

Ultimately, our research combines CFD simulations with real-time wind measurements to refine wind energy potential assessments in urban settings. By identifying localized wind acceleration zones, we provide a methodology for optimizing the placement of vertical-axis wind turbines, improving their efficiency in dense urban environments. This approach enables architects to integrate wind-harvesting solutions seamlessly into building design, advancing sustainable architecture towards energy autonomy.

Integrating aerofoils and BID devices represents a significant step towards rethinking architectural forms to optimize wind energy utilization. We propose a new design paradigm that prioritizes energy efficiency by leveraging urban wind corridors and optimizing building profiles. The methodology outlined in this study offers architects a strategic framework for designing buildings that consume less energy and contribute to renewable energy production, aligning with contemporary sustainability goals.

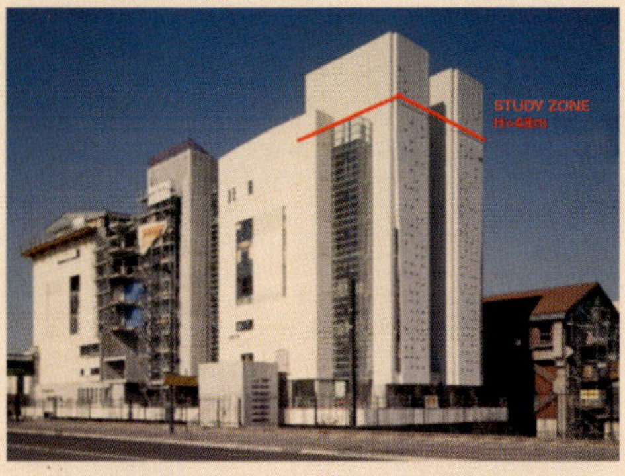

A building-block: the ENSA Paris-Val de Seine

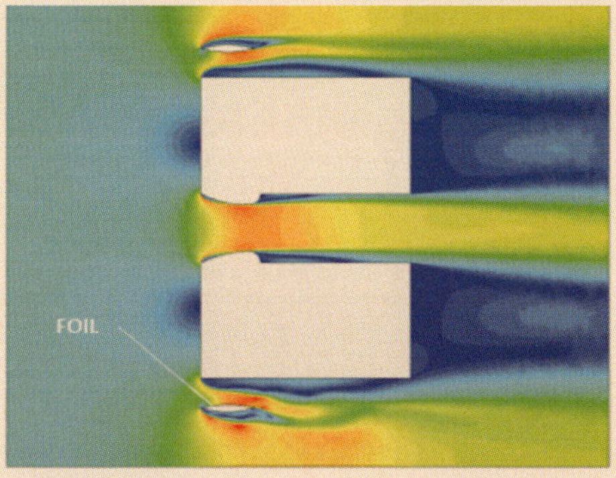

Aeraulic study in horizontal section with 2 parallel foils

Aeraulic study in the horizontal section of the building towers at 48 m high

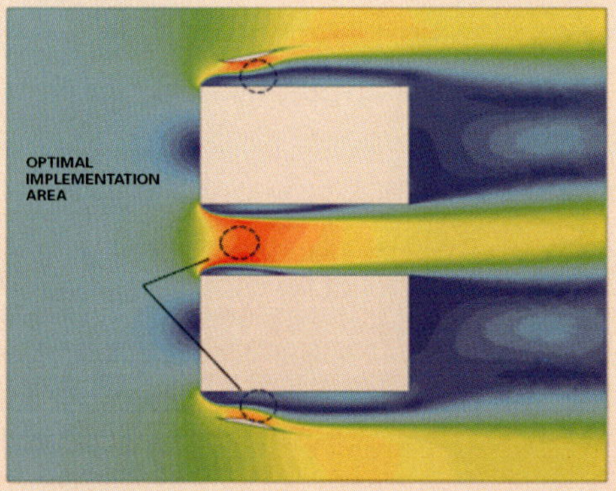

Aeraulic study in the horizontal section with two inclined foils

CFD Simulation of air movement around building towers with and without aerofoil

Aerofoil building locali-zation

Future research will focus on refining wind analysis techniques, developing adaptive façade systems, and exploring new materials that enhance energy efficiency. This research combines architectural innovation and advanced simulation technologies, establishing a foundation for the next generation of energy-positive buildings.

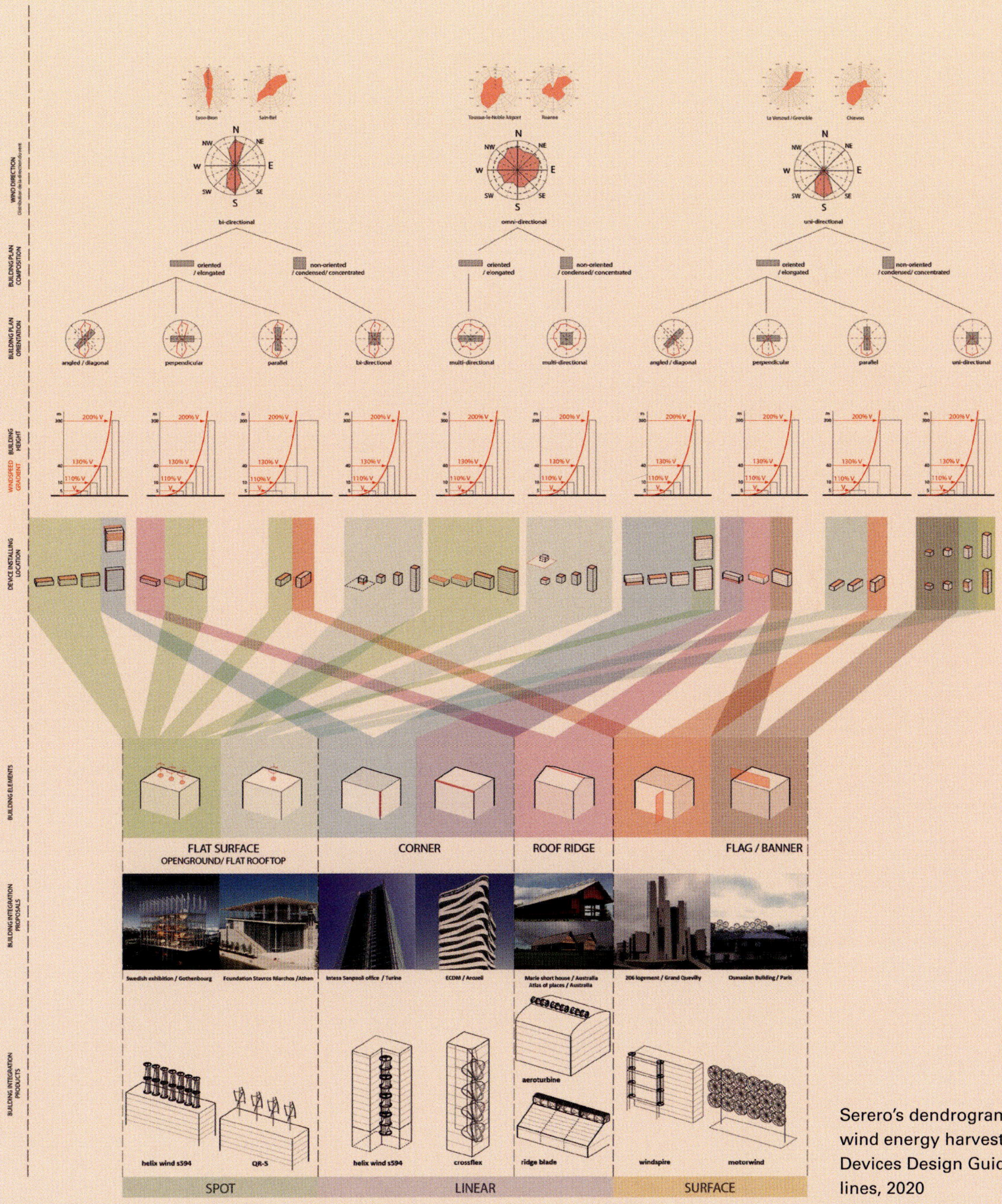

Serero's dendrogram for wind energy harvesting Devices Design Guidelines, 2020

Chapter 3

DATA ENERGY MATTER

Matter
From Informed Matter to Reused Matter

Present Past materials

Matter is explored in the project as an architectural device, to create a dialogue with the history of the place in some projects, or to create a physical or conceptual experience of the material in others, or even as an alternative construction method using reclaimed materials in a militant stance to preserve natural resources.

In many projects, the material dimension constitutes a powerful mode of dialogue with the past of the place. The choice of material drawn from local and historical resources makes it possible to forge links and open a relationship with adjacent or heritage buildings, either by similarity or by contrast with an existing condition. This strategy of dialogue is implemented in the project for the library of Saint-Hilaire-du-Harcouët, for the Museum of the Battle of Fromelles, and for the media library of Lezoux.

In the Saint-Hilaire library, a relentless search for stone in local granite quarries has made it possible to trace the origin of the unique pink granite with rust stains that were used in the 13th century in the listed church opposite the building.

For the Museum of the Battle of Fromelles, the geological stratifications of the land of the archaeological excavations and also of the trenches of the First World War are embedded in the walls of the museum by concrete of earth poured in successive layers.

In the Lezoux media library, the presence of a major site for the manufacture of Roman sigillata pottery from the 5th century BC has guided the use of fired bricks for the walls of the building, with the rewriting of stacking patterns, evoking both books on shelves and ventilation systems for the building, or acoustic perforation for the treatment of the workspaces of the media library.

In the projects of the Glass Loft in New York, and the reserves of the museums of Bayonne, the material is above all a conceptual device that offers a global experience of architecture, an atmosphere in which the project becomes the site of a physical and personal experience. Through experience, the material evokes a relationship with the place, the body and the site.

In the case of the Réserves de Bayonne, the entirely opaque volume of the storage space for works of art from museums in the Basque Country generates the opportunity for the generalization of a landscape envelope. Made of a hammered mirror-polished stainless steel skin, the building reflects the landscape of the heights of Bayonne in touches, like an impressionist painting by Claude Monet, transforming the reflection of the city and the sky into an atmospheric experience of color and landscape.

The glass loft is, above all, an exploration of the crystalline quality of Glass to qualify the relationship between intimate space and living space of this New York loft. Glass is not used for its transparency but rather for its structural and sculptural quality to define the interior architecture of an open living space.

The tree workshop addresses the question of matter through the reuse of building materials. The project is a virtuous and militant attempt to reuse structures and materials from other construction sites or from the demolition of replaced buildings to create a unique workspace in the heart of a garden block. The existing trees on the site have been incorporated into the project to enhance the interior spaces of the workshop and encourage the emergence of a living ecosystem in symbiosis with the users of the space.

Activating a Circular Economy of Material Resources

This essay, the result of research with D. Toubanos and D. Sipp, examines the transformation of architectural and urban practices in a context of increasing scarcity of material resources and global reflection on the sustainability of buildings and their life cycle. This crisis of primary resources is forcing architects to develop alternative methods of project design and construction based on the principle that the material resource is at the origin of the project's genesis, upstream of the architectural concept.

The work begins with the identification and rediscovery of the sources of materials and products available for architecture. The choice of materials, sometimes left in the hands of the builders, is once again becoming a fundamental element of the project for architects who define their specificities or their origin, in relation to their impact on the overall life cycle of the building.

Through a pedagogical experiment, this approach aims to explore the collection and referencing of these deposits. The geolocation of these resources using GPS technology allows for the continuous referencing and reuse of unused materials. This is the path of reuse in architecture. Materials from deconstruction, products from industrial or artisanal production, unsold goods or customer returns, the city, its businesses, and its users are full of potential resources that architects must explore in their production. The found object, diversion, reuse, and recycling are all strategies that contribute to the creativity and invention of the project, as well as to the development of an alternative materiality and image of the project.

This essay also discusses how to prototype structures and test them to evaluate the performance of unknown materials and assemblies. This is a complex mission where the scientific method offers a framework and an approach to study and discover these new potentials.

Open Physical and Virtual Territories

We are talking here about the migration of part of our physical environments to a virtual environment that is underway and will gradually push architects to intervene, such as in cities, rural environments and social networks.

Designing architecture means drawing on our most personal emotional experiences, but what will become of this architecture when architects have spent more time in immersive worlds than in physical spaces?

In the first reality, most of our experiences and emotions are also generated by social relationships, by our encounters or interactions. That is why we are working on an augmented reality design model and not on that of a closed virtual world.

In these closed virtual worlds, as opposed to our physical reality, we will never be surprised by an unexpected bird song in this environment, a hubbub that reveals the complexity of the occupation of spaces, by the freshness of the air in an outdoor space, the smell of moss as we cross a space, the warmth of a ray of sunshine or a breath on our face. Imagine a world where the simple 3, 4, or 5 dimensions are replaced by a complete hybrid reality, a multidimensional reality in time and space, including all users.

For us, it is a complete hybrid world where primary reality and all its physical, climatic, biological and chemical components combine and amplify their experience with each other. Building with heterogeneous and variable materials, biomaterials, combined with industrial materials, allowing for a hybridity of skill and materiality. Faced with the beauty and complexity of our reality, the creation of an augmented reality model is only an intermediate stage in the experience of a reality recomposed by man.

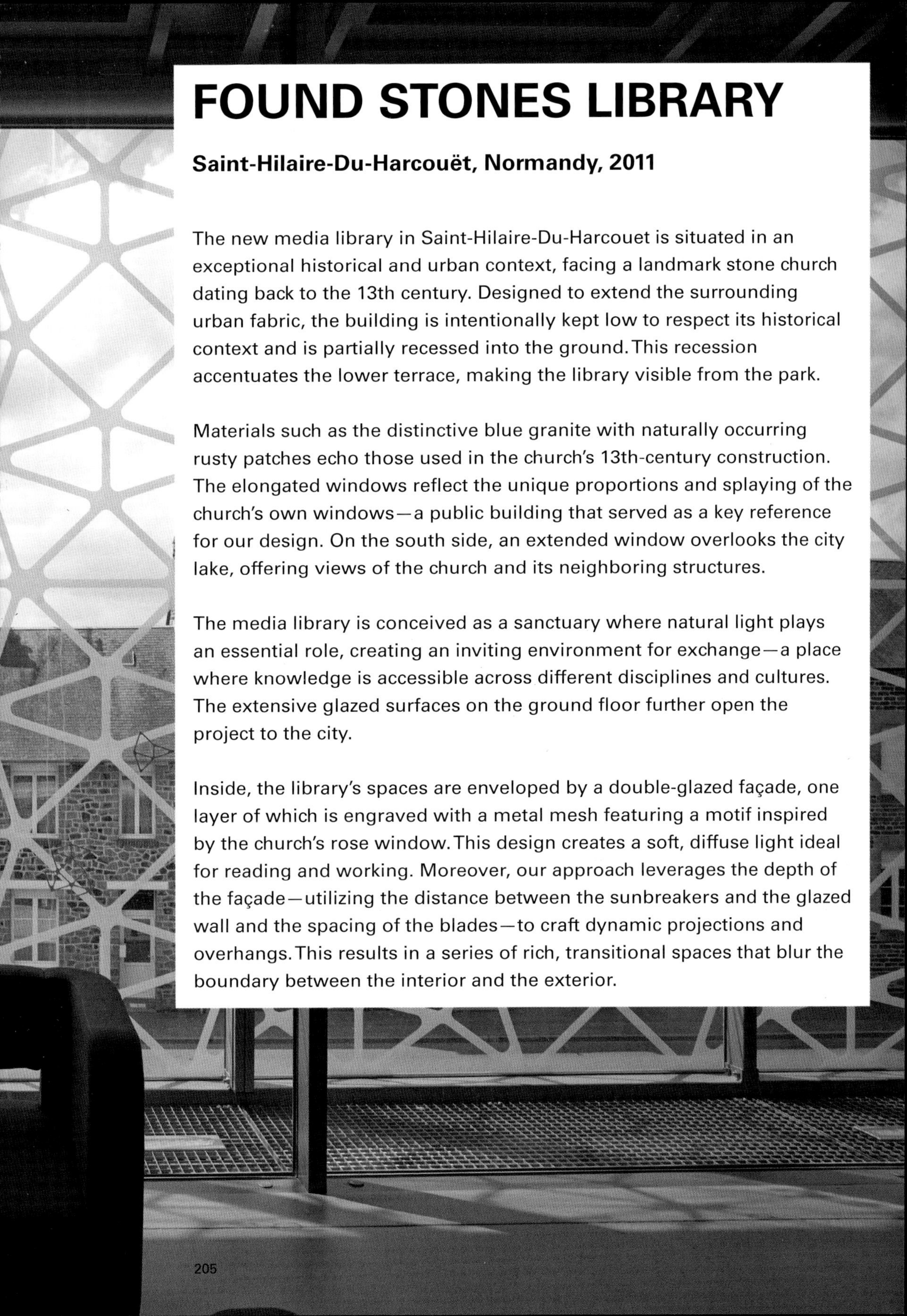

FOUND STONES LIBRARY

Saint-Hilaire-Du-Harcouët, Normandy, 2011

The new media library in Saint-Hilaire-Du-Harcouet is situated in an exceptional historical and urban context, facing a landmark stone church dating back to the 13th century. Designed to extend the surrounding urban fabric, the building is intentionally kept low to respect its historical context and is partially recessed into the ground. This recession accentuates the lower terrace, making the library visible from the park.

Materials such as the distinctive blue granite with naturally occurring rusty patches echo those used in the church's 13th-century construction. The elongated windows reflect the unique proportions and splaying of the church's own windows—a public building that served as a key reference for our design. On the south side, an extended window overlooks the city lake, offering views of the church and its neighboring structures.

The media library is conceived as a sanctuary where natural light plays an essential role, creating an inviting environment for exchange—a place where knowledge is accessible across different disciplines and cultures. The extensive glazed surfaces on the ground floor further open the project to the city.

Inside, the library's spaces are enveloped by a double-glazed façade, one layer of which is engraved with a metal mesh featuring a motif inspired by the church's rose window. This design creates a soft, diffuse light ideal for reading and working. Moreover, our approach leverages the depth of the façade—utilizing the distance between the sunbreakers and the glazed wall and the spacing of the blades—to craft dynamic projections and overhangs. This results in a series of rich, transitional spaces that blur the boundary between the interior and the exterior.

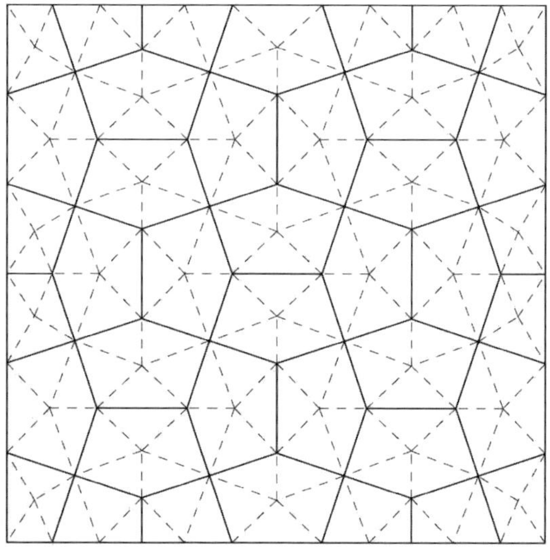

Glass wall patterns inspired by
13th century cathedral rose window

MÉDIATHÈQUE MARIN-MARIE

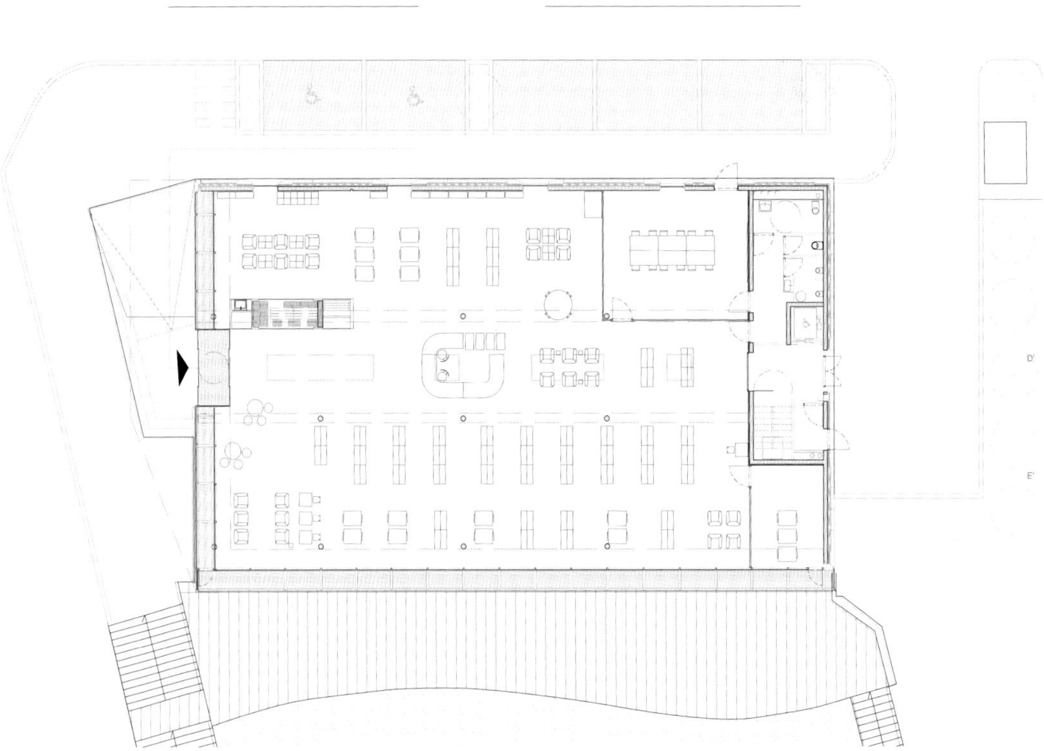

Ground floor plan

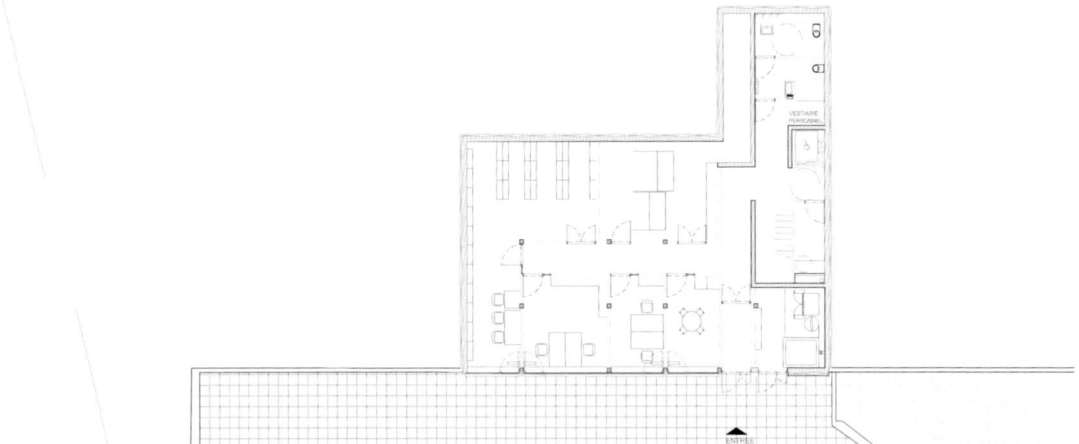

Garden level floor plan

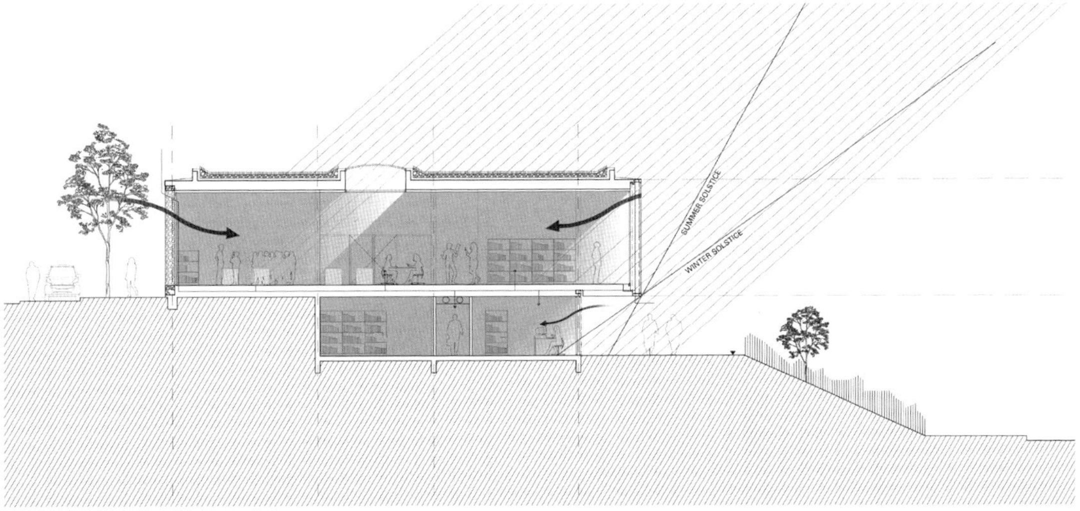

Landscape section of building

BATTLE OF FROMELLES MUSEUM

Lille, 2016

On 19 July 1916, several battalions of the 5th Australian Division crossed the German front lines but were surrounded until the battle's end on 20 July. The wounded were treated, prisoners interrogated, and the dead buried in five mass graves. In 1999, a local historian reported the site, leading to an exhumation in 2009 by the Commonwealth War Graves Commission, which uncovered 250 bodies. A cemetery was created in Fromelles to provide them with a dignified resting place, where a museum dedicated to the battle is located.

The Fromelles Museum's architecture aims to connect heaven and earth, creating a bond between visitors and the soldiers' resting place. The design integrates with the site's natural slope, with the main exhibition hall emerging from a mound of vegetation. It utilizes raw concrete in earthy tones to evoke geological layers and the burial of soldiers.

The museum's low profile complements the adjacent military cemetery, with two main visual axes: one directing attention to the cemetery and the other to the prominent Fromelles church. At the center lies an octagonal exhibition hall that reflects military geometry and the sacred nature of the nearby church.

Natural light is crucial throughout the museum, with reception areas illuminated by light shafts. The exhibition area begins in darkness and gradually brightens. In the evening, outdoor lighting enhances the commemorative spaces.

The museography centers on the Battle of Fromelles within broader narratives. Visitors walk along an elevated pathway reminiscent of archaeological digs, which transforms into a symbolic trench. This culminates in the "War Room", a circular space featuring a video installation on the battle.

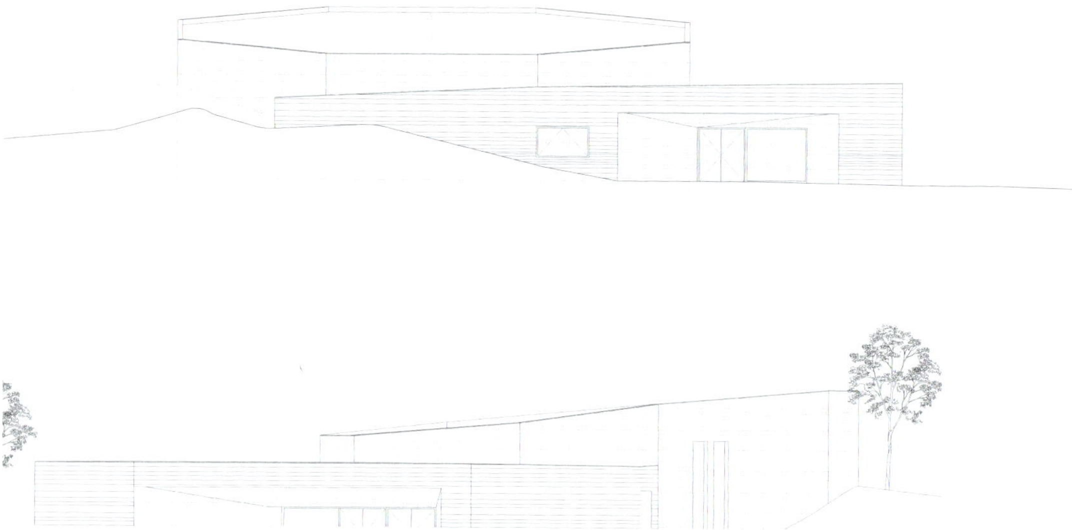

Elevations of striated concrete museum walls

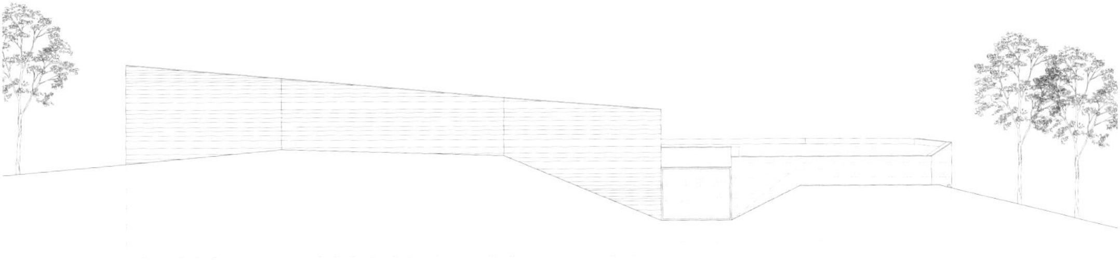

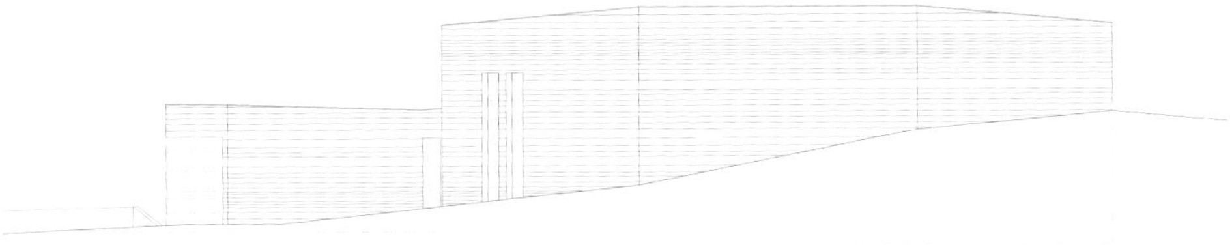

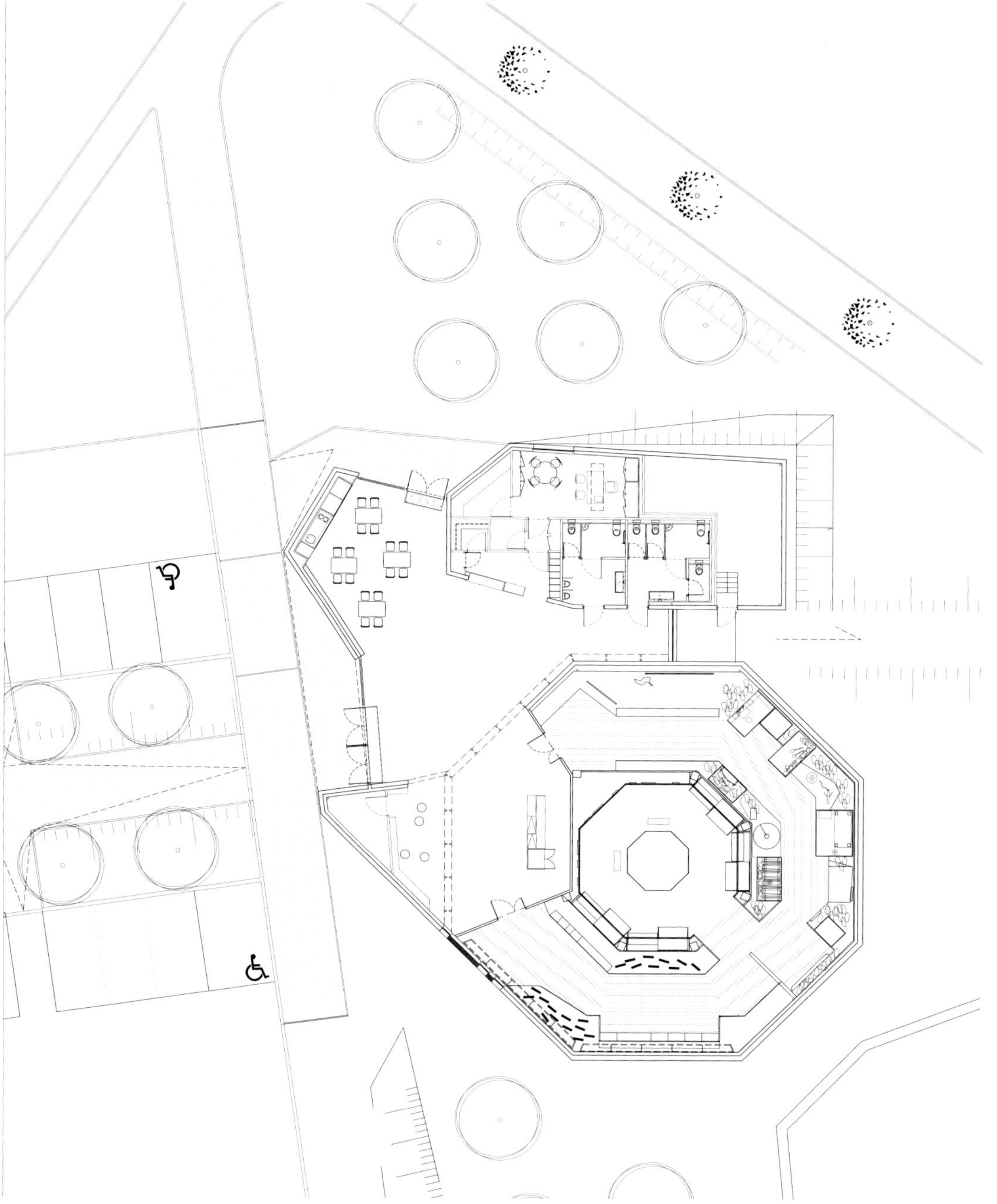

Ground floor plan

Connecting museum lobby to the battle fields remaining building landmarks

READING IN A GARDEN

Lezoux Library
Auvergne Alpes, 2018

The new media library in Lezoux occupies a site that serves as a breathing space within the city, offering views of the surrounding landscape. In response to this unique setting, we proposed a pavilion-style structure—a welcoming hub where books, multimedia resources, and activities are accessible amidst a lush garden. Its transparent design blurs the site's boundaries, transforming the building into a genuine urban showcase.

The structure features an openwork roof that shelters a transparent volume housing consultation areas and a reading room. Supported by steel pillars, the roof leans against two opaque concrete masses that act as bracing elements. Its slender silhouette and perforated design evoke the ambiance of forest undergrowth. Integrated roller blinds enhance sun protection and visual comfort, creating an inviting environment for reading and work.

Inside, the media library is envisioned as a multipurpose space with fluid boundaries, fostering interaction across generations. The space exudes transparency and facilitates access to its collections. It transitions smoothly from a grand reception area to quieter zones ideal for concentration and individual work, leading to intimate, staff-dedicated areas. Two 'porous' rooms opposite the reading room provide quiet workspaces that allow users to isolate themselves or connect with the exterior via the adjacent reading terrace.

A series of distinctive gardens surround the building, enlivening the interior spaces of the media library. Birch trees have been planted throughout the site to accentuate these outdoor spaces. Notably, the "Tree of Wisdom," with its luminous foliage and parchment-like white bark, echoes the building's columns.

The forecourt is accessed via andesite lava stone paving, while the surrounding land is sculpted into mounds covered with diverse meadowland. These mounds support an open-air theatre that opens onto the external stage of the multipurpose hall on the west side, complemented by wooden terraces that extend the library's activities outdoors.

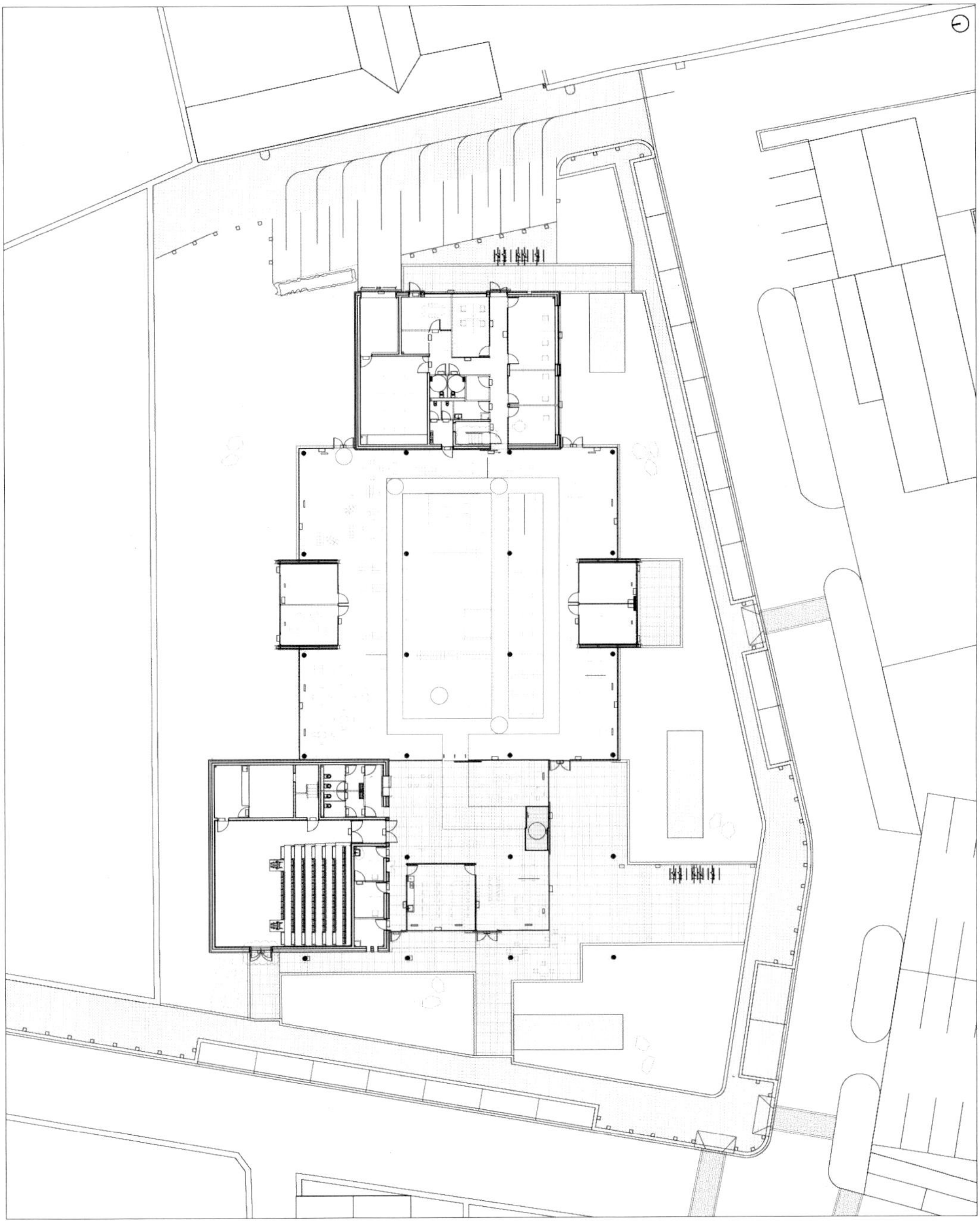

Ground floor plan

SECTION AA

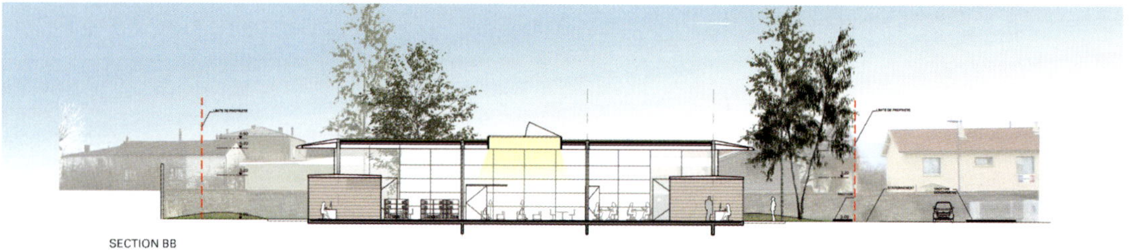

SECTION BB

Building and landscape sections

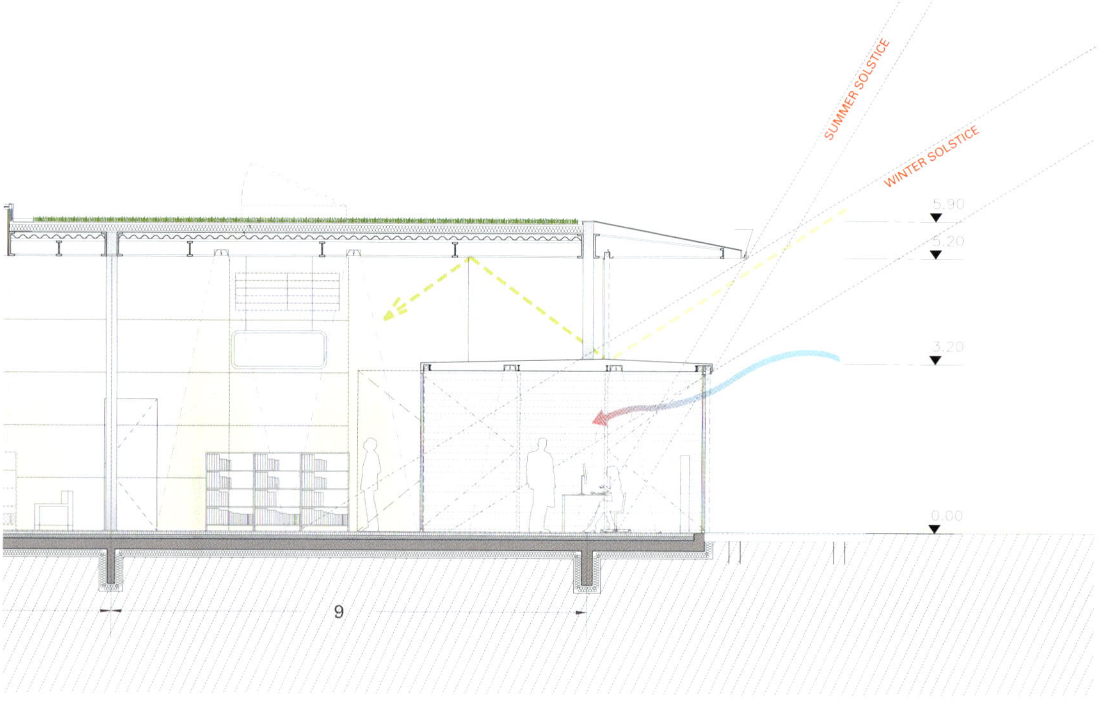

Media center bioclimatic section

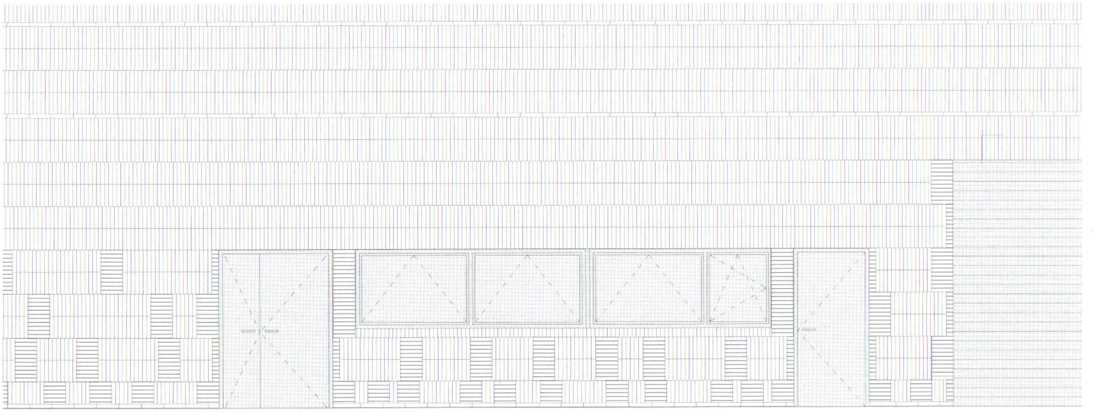

Brick structural, acoustical, and symbolic patterns

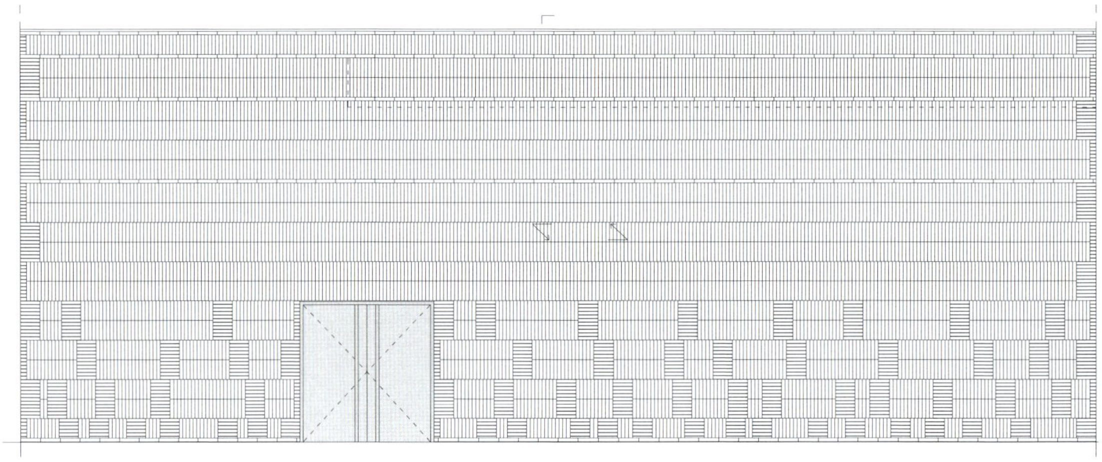

West elevation of auditorium "solid" block

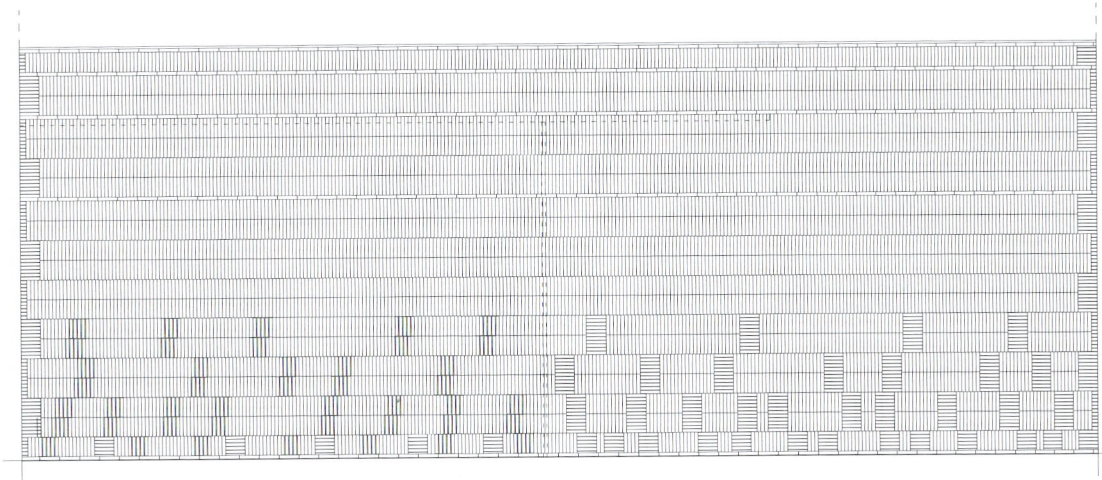

East elevation of auditorium "solid" block

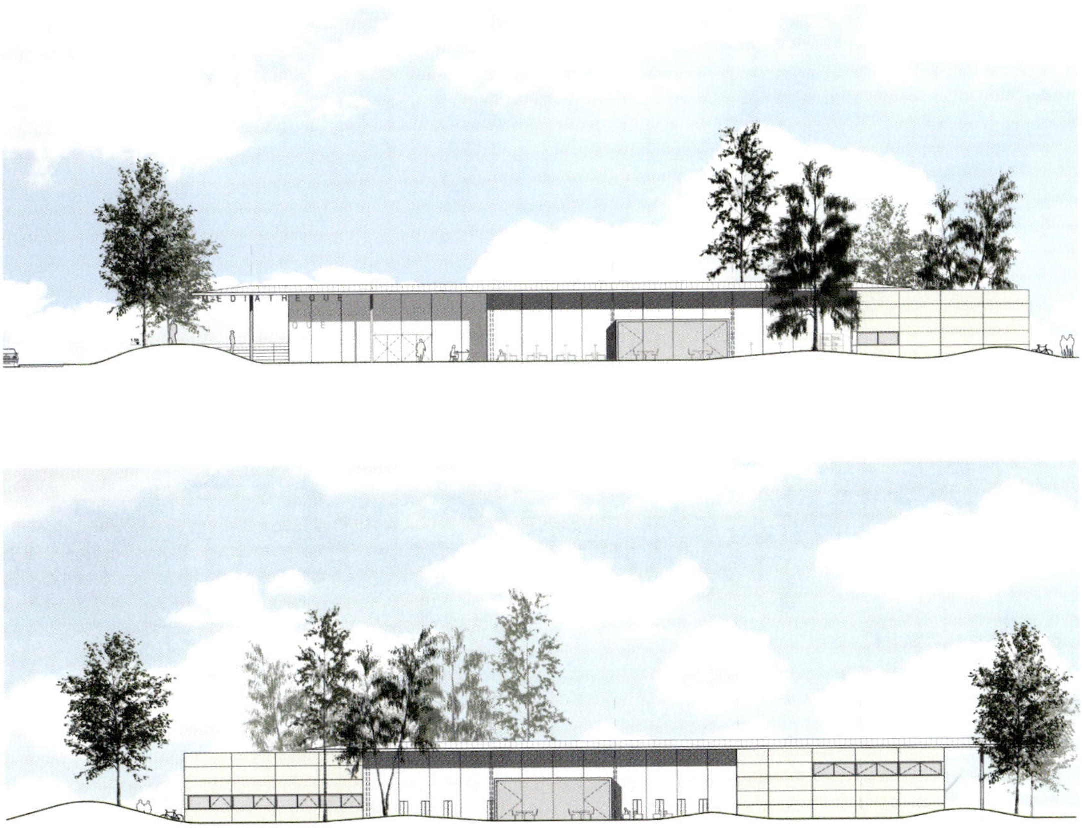

General south and north elevations

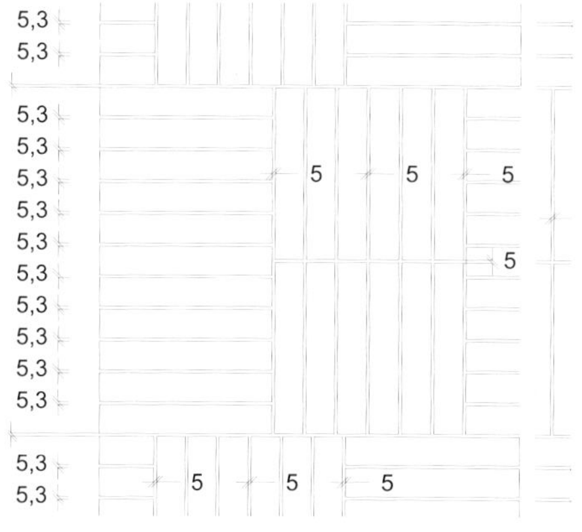

DETAIL 1: BRICK LAYOUT SIZES IN MILLIMETERS
ESH:1:5

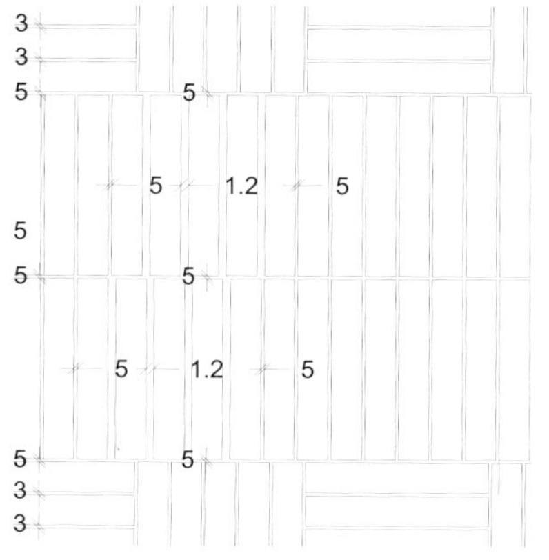

DETAIL 2: BRICK LAYOUT ON EDGE DIMENSIONS
IN MILLIMETERS ESH:1:5

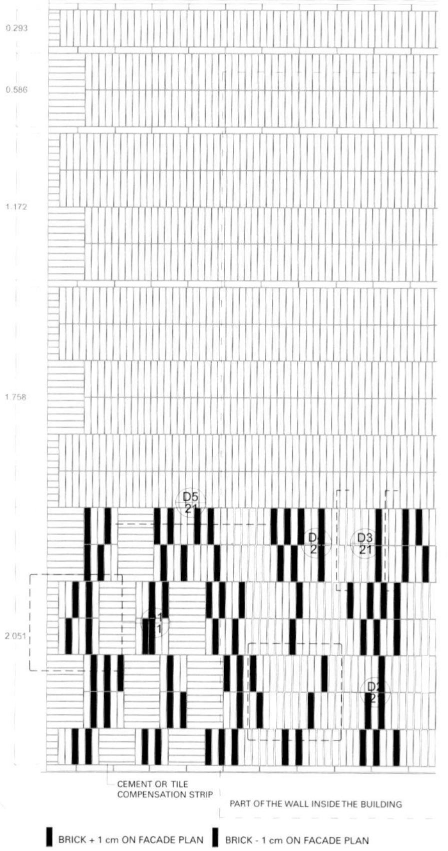

SECTION LAYOUT OPTION 01
ESH:1:20

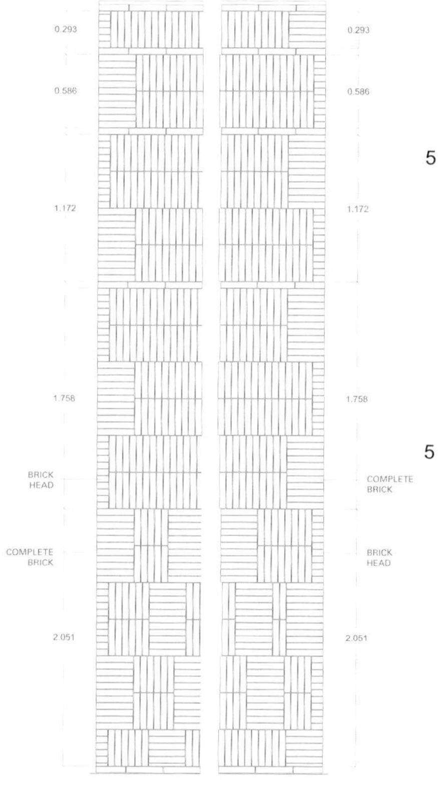

ELEVATION LAYOUT OPTION ANGLES
ESH:1:20

Brick assembly patterns

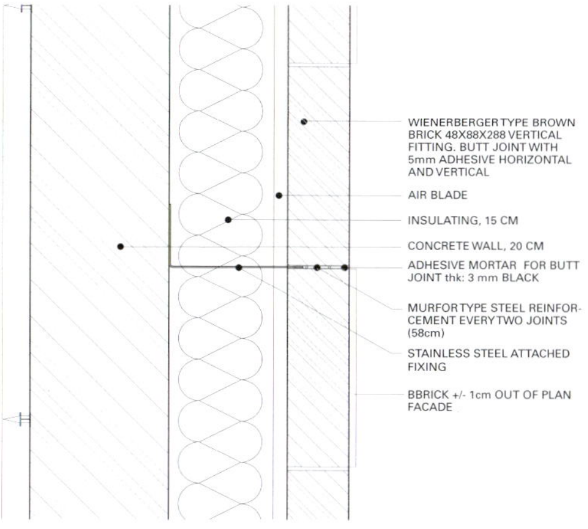

SECTION DETAIL 3
ECH:1:5

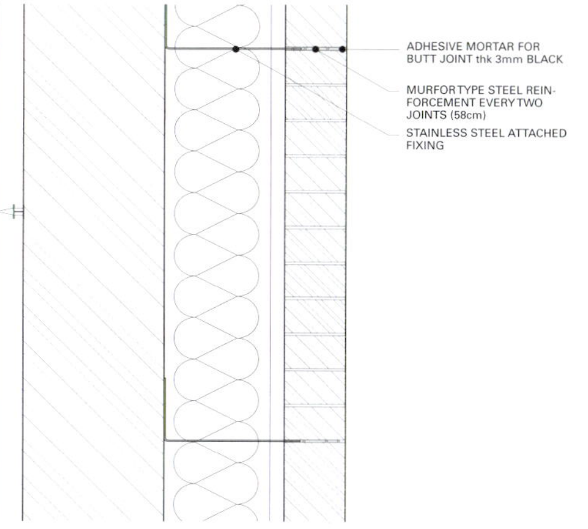

SECTION DETAIL 4
ECH:1:5

Sismic assembly detail for brick walls

Prototyping brick walls

Private working space terraces

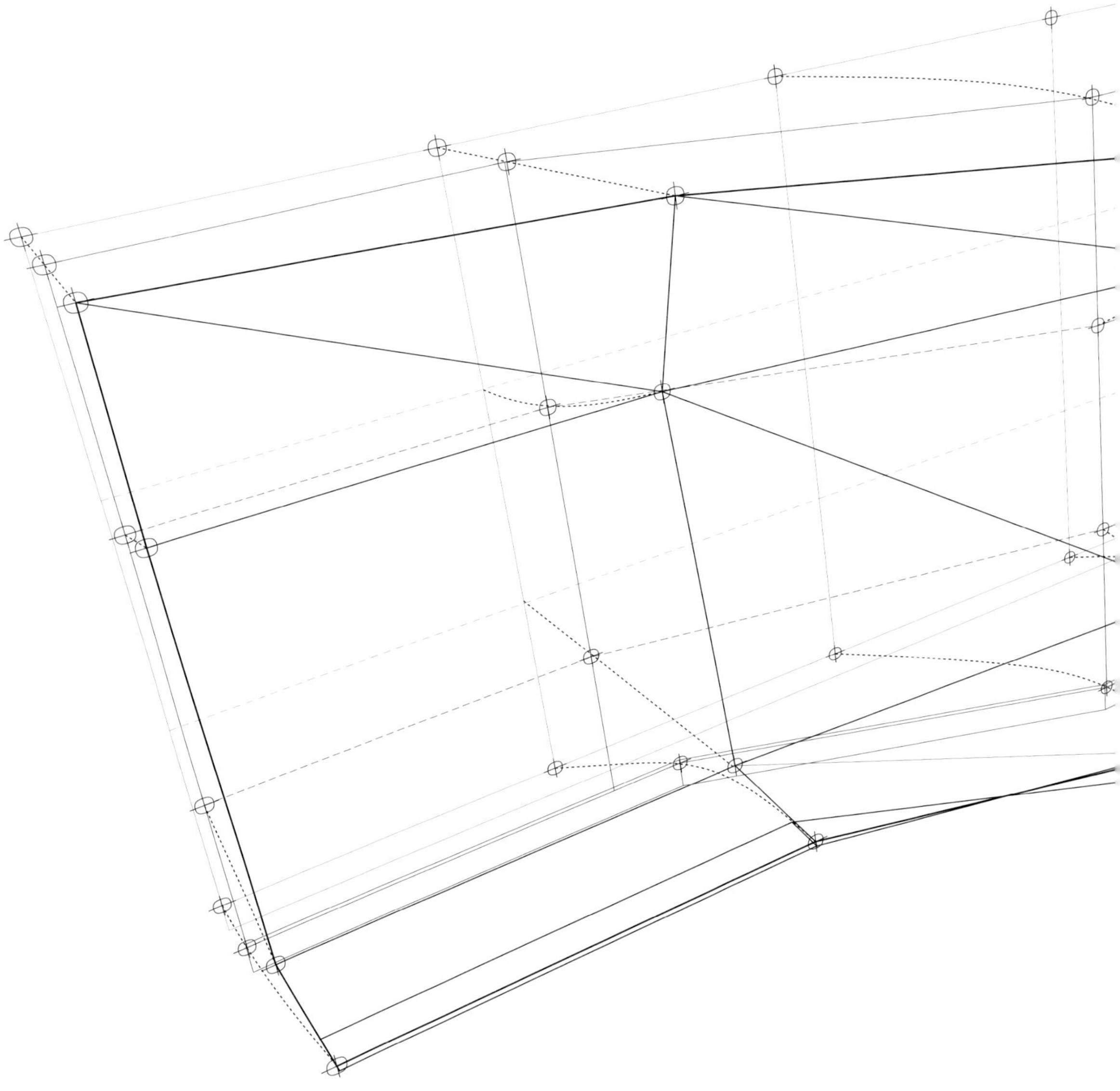

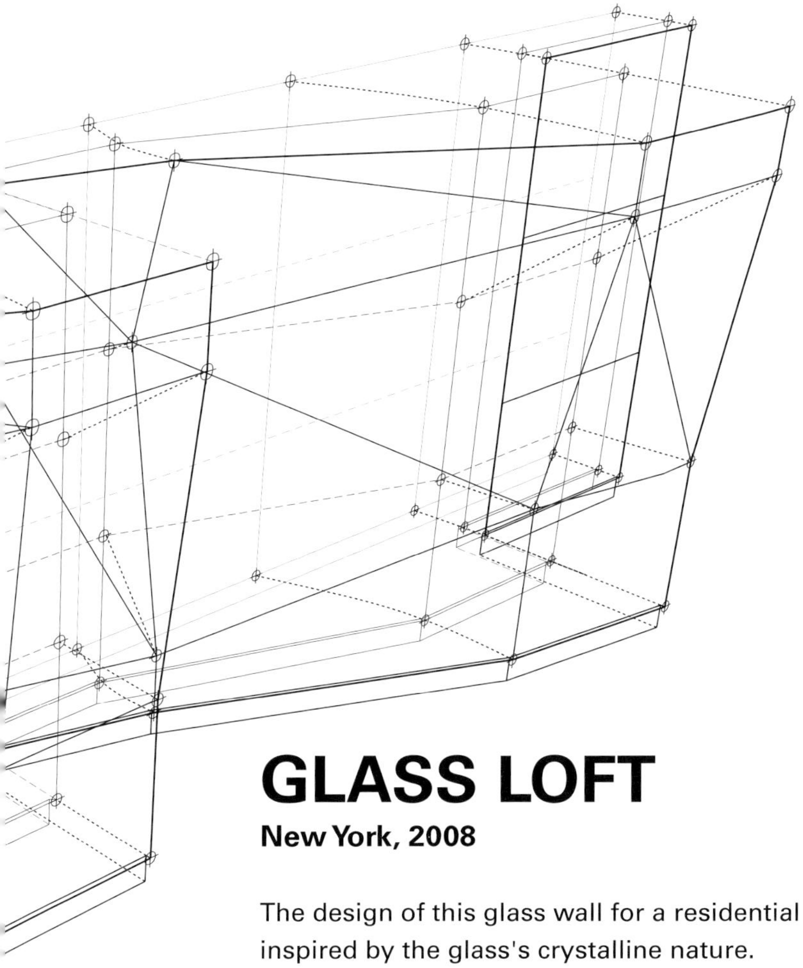

GLASS LOFT

New York, 2008

The design of this glass wall for a residential loft in New York was inspired by the glass's crystalline nature.

The wall oscillates within the loft, delineating living areas, sleeping quarters, and service zones while regulating both views and access between them.

Its structure interweaves with the building's existing wooden framework—sometimes anchored to it, sometimes cantilevered beyond it—creating a space of intense ambiguity between a rational structure and a free-form envelope.

The movement of the human wall'snformed the transformation of the wall’s geometry as it traverses space and interacts with its boundaries.

By inflecting the vector through which the body crosses the wall, the surface gains thickness and effectively doubles itself, thereby providing the necessary privacy for the two bedrooms behind it.

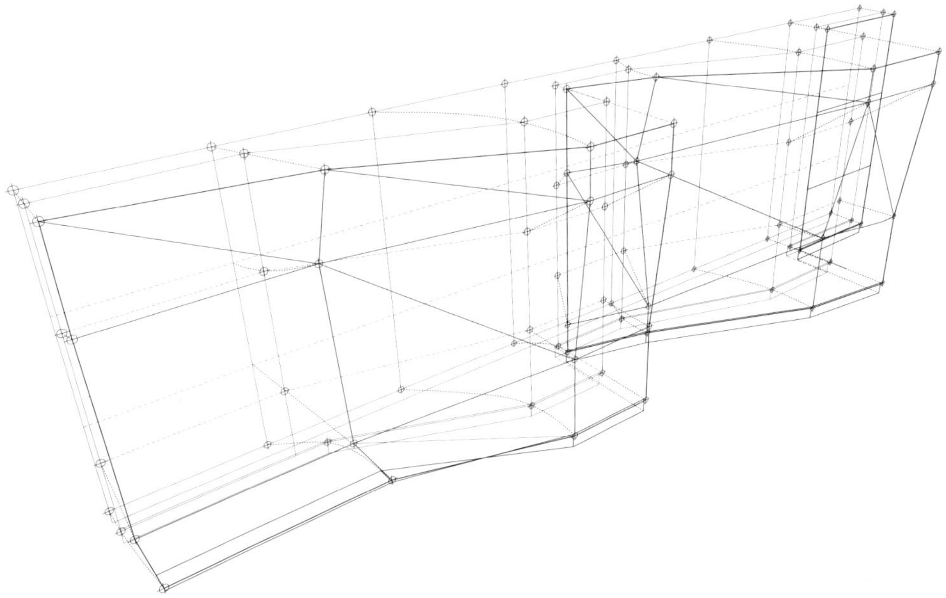

Geometric manipulation of glass wall to integrate living room and bedroom access

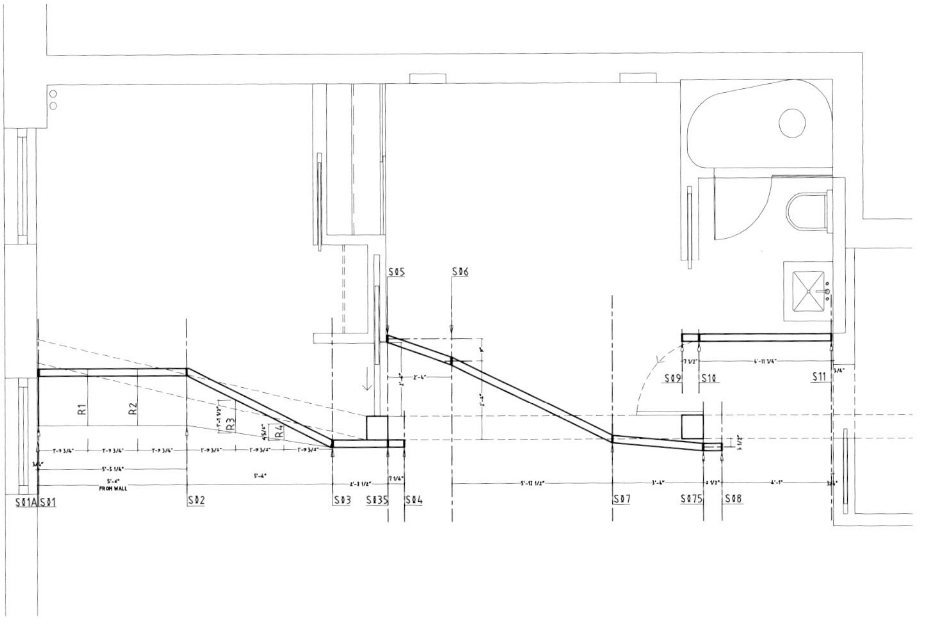

Detail plan of glass wall in appartement

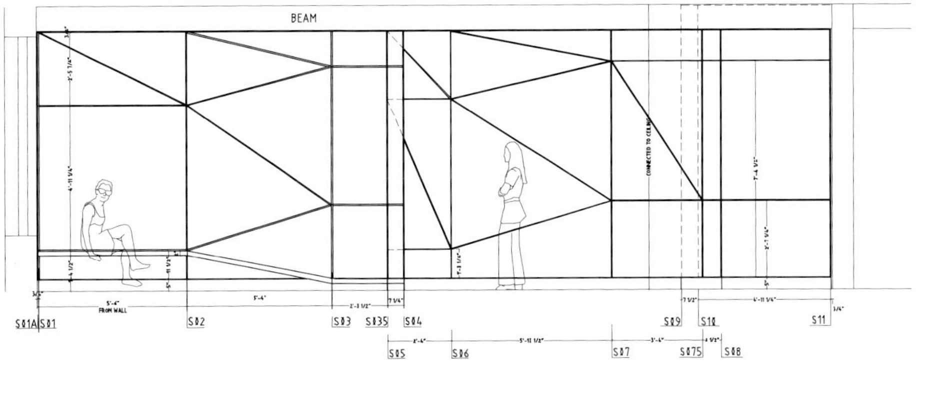

Gass wall elevation

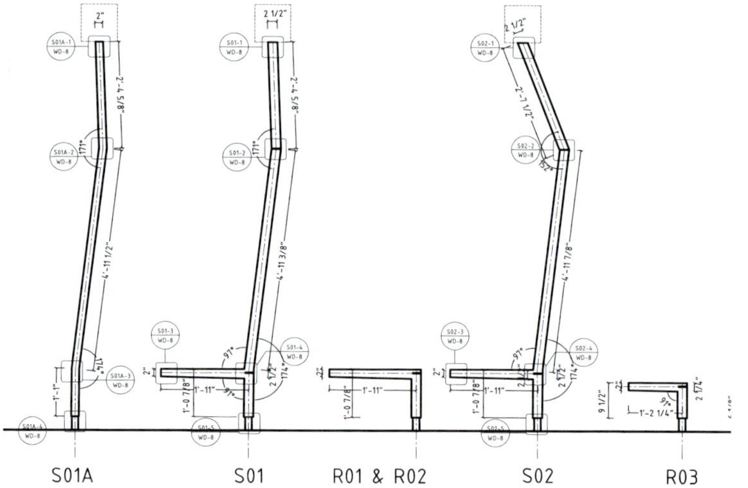

Variable glass wall sections

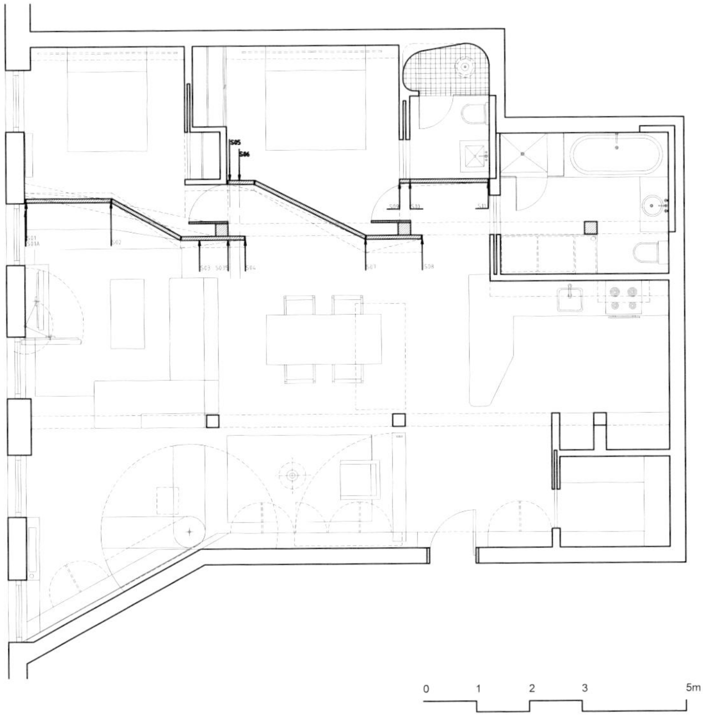

Loft floor plan

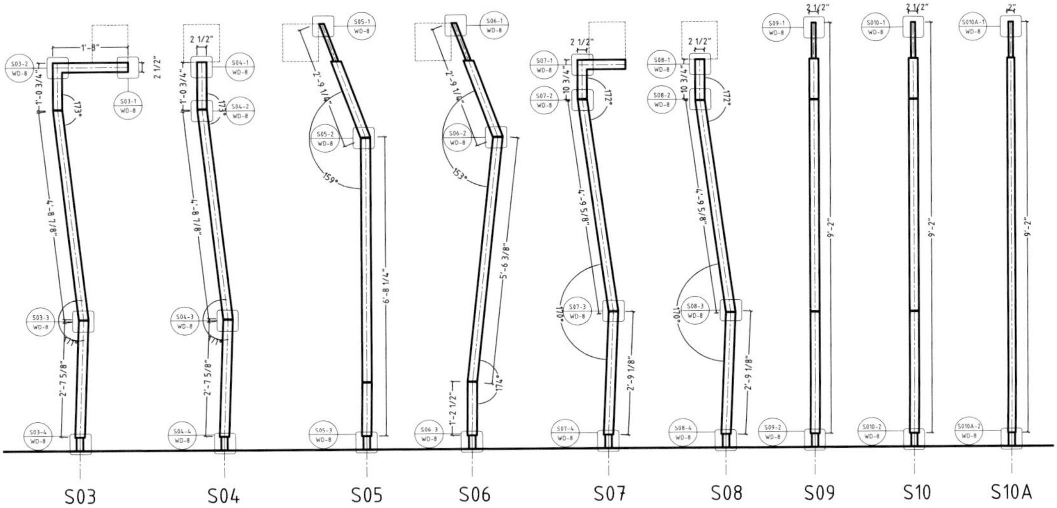

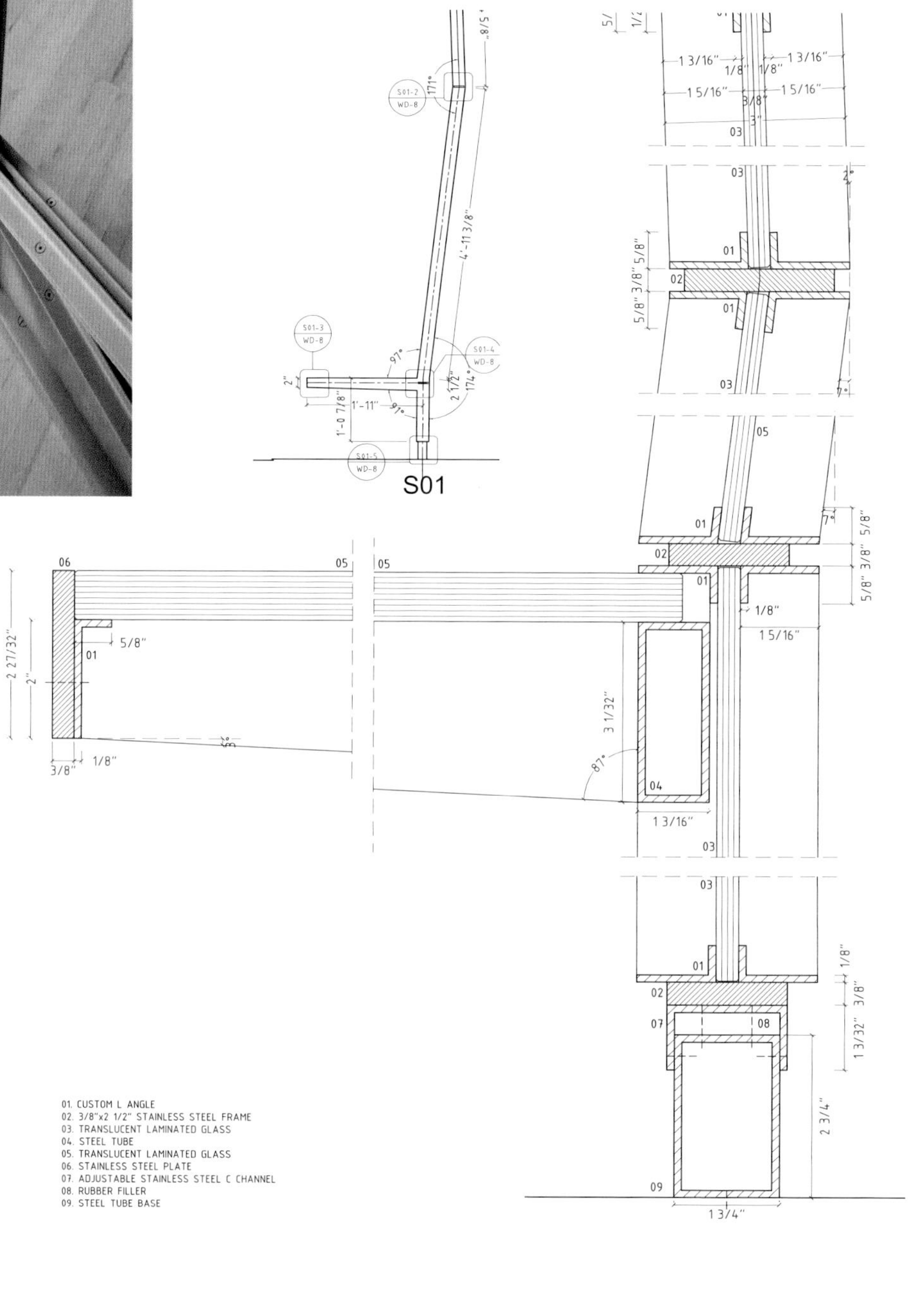

Steel and glass joinery details

Bluestone bath and shower bathroom

BASQUE COUNTRY MUSEUMS STORAGE FACILITY

Bayonne, 2022

The project to build shared museum storage facilities in Bayonne reflects the city's commitment to preserving its artistic and cultural heritage. These storage facilities enable precise control over the conservation conditions of each object, tailored to its typology, origin, and current state.

The design divides the storage areas into smaller cells, thereby increasing the versatility of the spaces and improving the organization of the stored objects. This minimalist building derives its overall layout from an internal organization based on stacking storage spaces.

The Bayonne Museum's storage facility design emerges from an exploration of architectural materiality intended to evoke both permanence—the enduring nature of objects—and the fleeting quality of the moment. The façade features cladding composed of large, mirror-polished steel panels produced by digitally stamping stainless steel sheets.

The folds in the sheet metal evoke the traditional craftsmanship of metalwork—reminiscent of hammered metal or the textured surfaces of bronze sculptures. This technique stiffens the façade panels and creates free-form patterns with seamless continuity between them.

On a contextual scale, this design feature offers a tangible experience of light and the surrounding environment. Much like the work of the impressionist painter Monet, the reflections on these surfaces capture an ephemeral moment. Rather than simply storing objects, the museum reserves are envisioned as active witnesses to the passage of time, capturing every fleeting second with intensity, with an architectural envelope whose materiality and geometry uniquely encapsulate these transient moments.

Four city-museums storage and archive space, Bayonne

Plan of artwork stacks and ateliers

Research on steel reflecting patterns

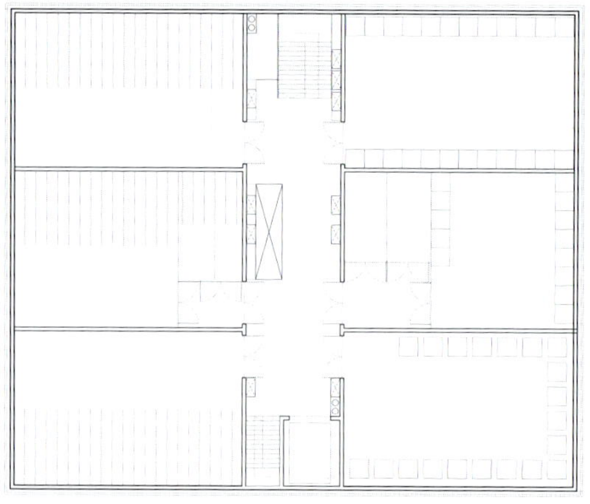

Floor plan
R+1 +4.00

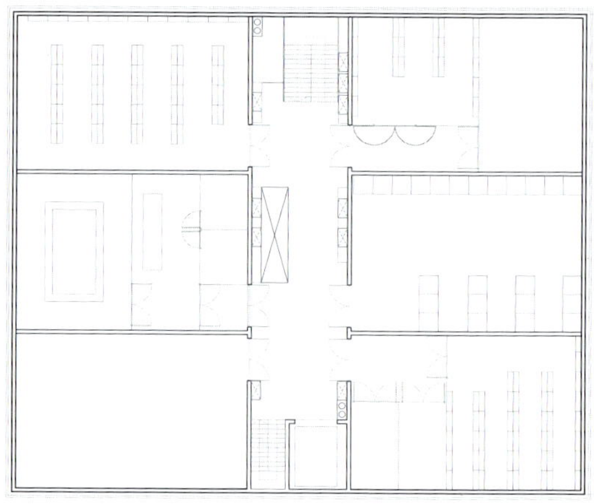

Floor plan
R+2 +8.00

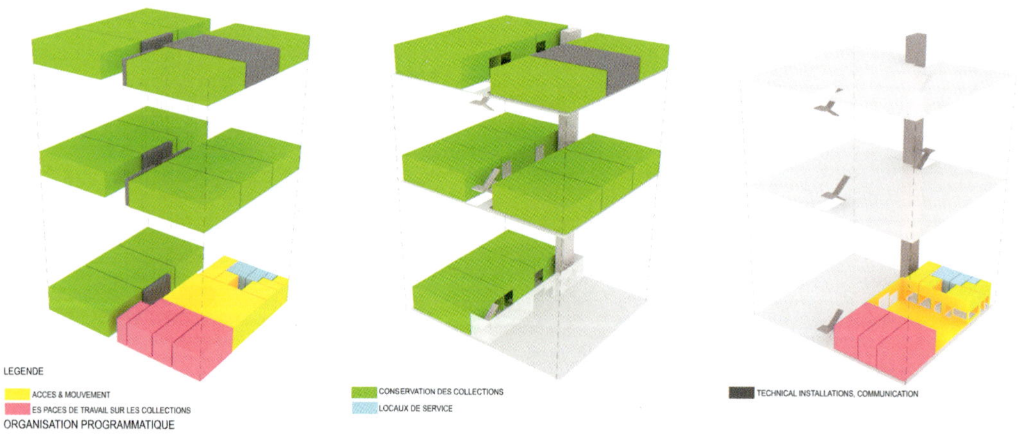

Programmatic assembly in a monolith organisation

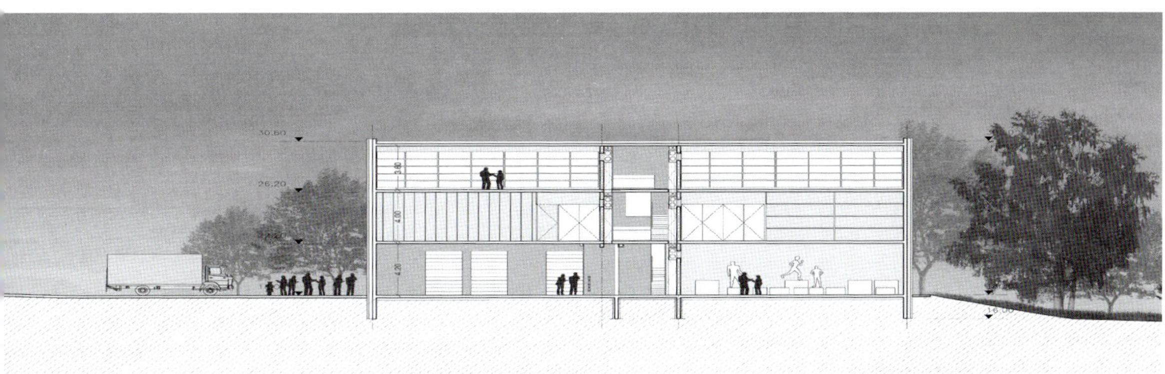

Section of building and stacks

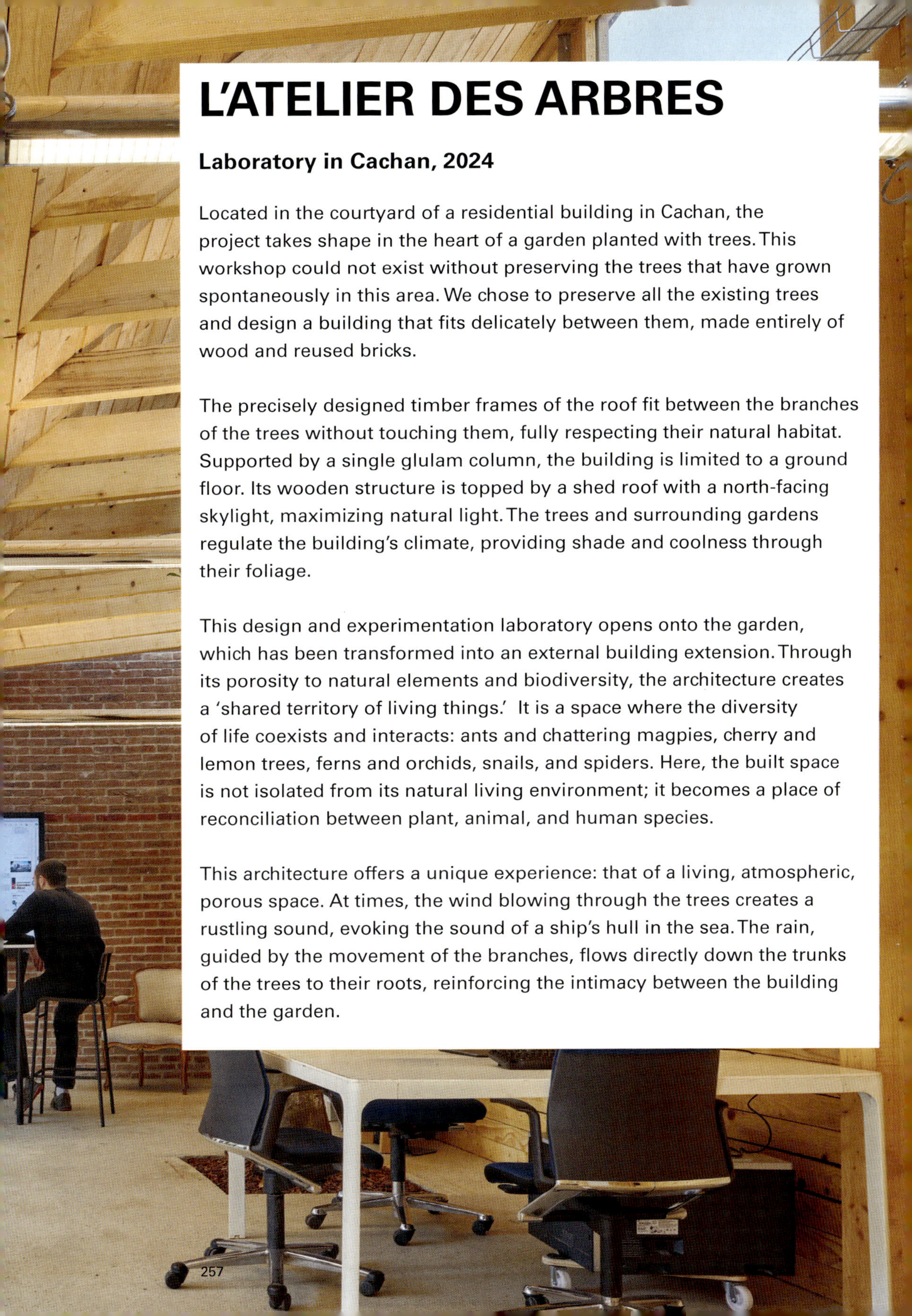

L'ATELIER DES ARBRES

Laboratory in Cachan, 2024

Located in the courtyard of a residential building in Cachan, the project takes shape in the heart of a garden planted with trees. This workshop could not exist without preserving the trees that have grown spontaneously in this area. We chose to preserve all the existing trees and design a building that fits delicately between them, made entirely of wood and reused bricks.

The precisely designed timber frames of the roof fit between the branches of the trees without touching them, fully respecting their natural habitat. Supported by a single glulam column, the building is limited to a ground floor. Its wooden structure is topped by a shed roof with a north-facing skylight, maximizing natural light. The trees and surrounding gardens regulate the building's climate, providing shade and coolness through their foliage.

This design and experimentation laboratory opens onto the garden, which has been transformed into an external building extension. Through its porosity to natural elements and biodiversity, the architecture creates a 'shared territory of living things.' It is a space where the diversity of life coexists and interacts: ants and chattering magpies, cherry and lemon trees, ferns and orchids, snails, and spiders. Here, the built space is not isolated from its natural living environment; it becomes a place of reconciliation between plant, animal, and human species.

This architecture offers a unique experience: that of a living, atmospheric, porous space. At times, the wind blowing through the trees creates a rustling sound, evoking the sound of a ship's hull in the sea. The rain, guided by the movement of the branches, flows directly down the trunks of the trees to their roots, reinforcing the intimacy between the building and the garden.

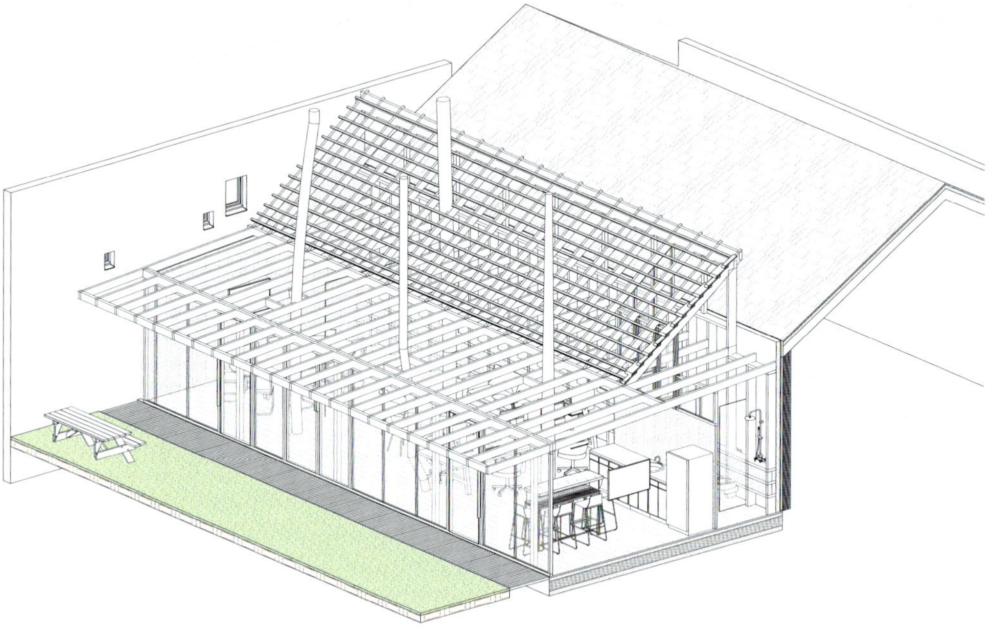

Programmatic assembly
in a monolith organization

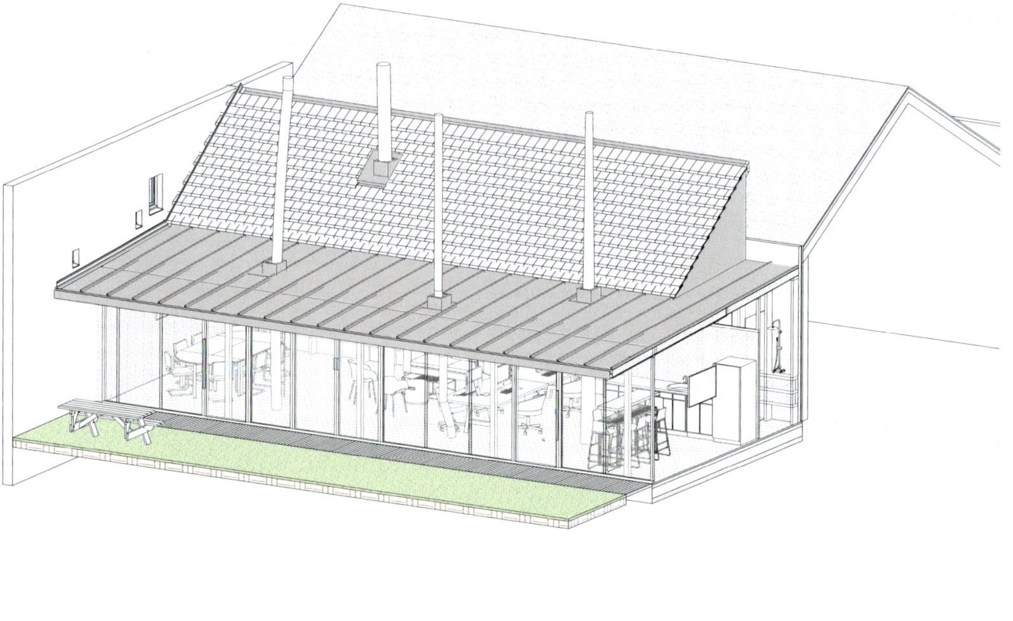

Axonometric view of structure
and trees intertwining

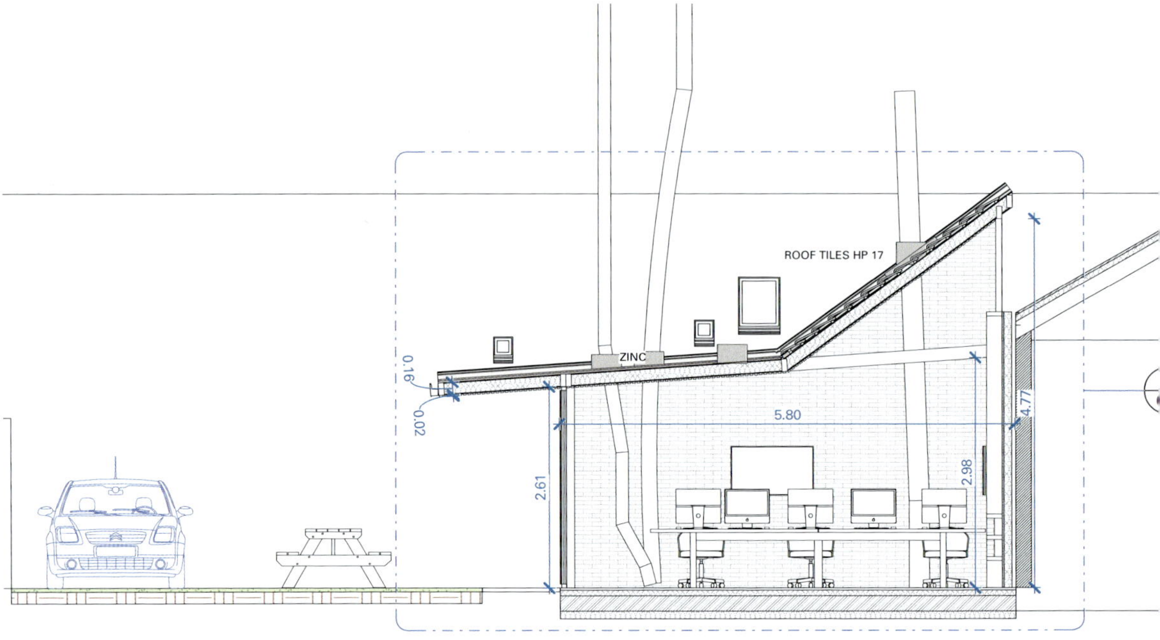

Wood carptentry and roof section

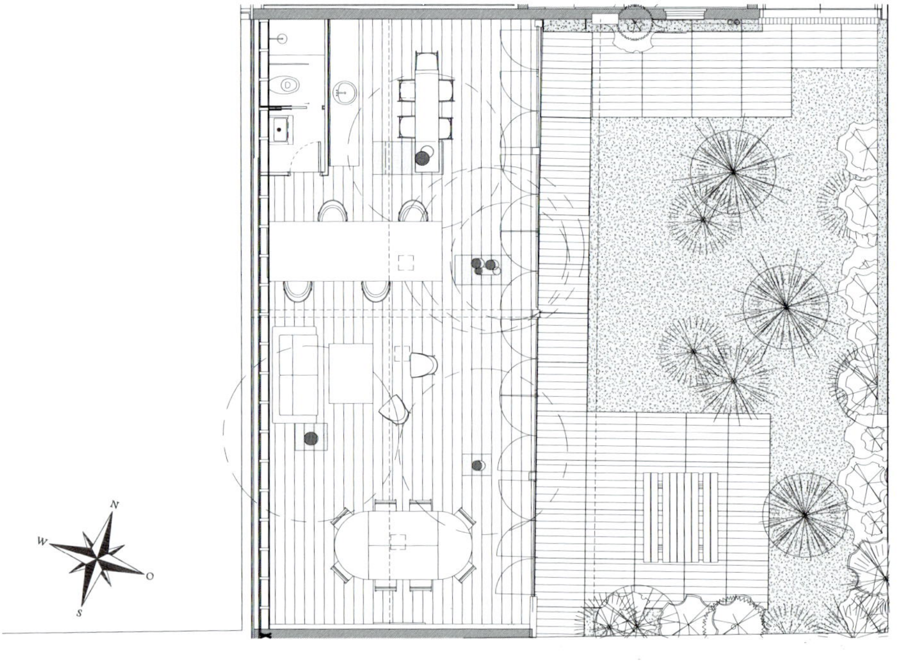

Ground floor plan

Generative AI for Architectural Design in Reuse Methodology and Experimentation at Scale 1

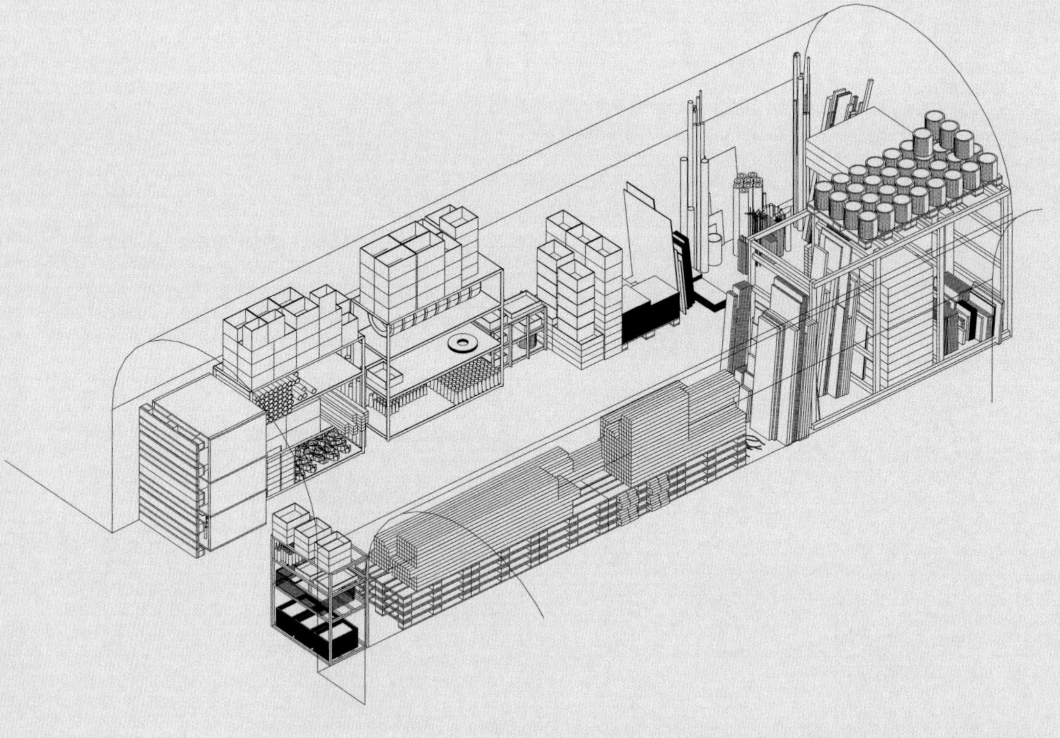

1_Axonometric view of ressouce center organised by ENSAPVS students

This research presents an alternative methodology for architectural design using reused construction materials, by assembling a pool of materials identified using text2image-type generative AI. This method aims to enumerate a set of heterogeneous material resources and a model for recommending assembly.

Research Context

The cumulative impact of construction activity is estimated at 45 to 55% of the waste produced worldwide[1]. The responsibility of architects, engineers, and clients alike is indisputable, but the construction of our environment and its transformation also offer potential for new solutions. It opens up a reflection on the transformation of our lifestyles and our impact on the earth. Inhabiting our territory should no longer mean building new "artificial surfaces"[2] to replace the natural ecosystem. In the long term, architecture should no longer be about building by adding and extracting resources, but rather about reversing the cycle of matter by deconstructing buildings and rebuilding our natural resources. This inverted vision of human activity is now at a tipping point. The convergence of analyses of the acceleration of climate change and its impact on living ecosystems is mature and commits us to deploy all our efforts to accelerate this paradigm shift.

1 Survey "Waste and Rubble Produced by Construction Activity" in 2014", SOeS.

2 The Net Zero Artificialization (ZAN) approach is an objective set for 2050. Enshrined in 2018 by the Biodiversity Plan and then in 2020 by the Citizens' Climate Convention, this approach consists of minimizing the expansion of cities by limiting construction on natural or agricultural spaces and compensating for urbanization by giving more space to nature in the city.

This research examines the transformation of architectural and urban practices in a context of increasing material scarcity and global reflection on the sustainability of buildings and their life cycles. This primary resource crisis forces architects to develop alternative methods of project design and construction based on the premise that the material resource is at the origin of the project's genesis, upstream of the architectural concept.

This material approach to the project forces us to rethink the notion of design, or rather of "Designing the design" as the sociologist Edgar Morin evokes it in volume 3 of his quasi-encyclopedic work

The Method[3], which suggests taking a step back from the objective. Thinking about the method of architecture means thinking about design rather than the simple end purpose itself: architecture.

3 Morin, Edgar. *The Method ,Vol 3 ; la connaissance de la connaissance*, Seuil, 1981, 416 p.

The question of research in "design" mobilizes a specific epistemological framework that is different from the classical framework of scientific research, as described by H. Simons[4], that is to say, the mobilization of know-how for the production of artifacts, rather than a traditional corpus of research. It is a domain where the scientific method offers a working framework and an approach for studying and opening up new potentials for "design".

4 Simon, Herbert A.; *The Sciences of the Artificial*, 3rd Edition, The MIT Press Paperback, ISBN: 9780262691918, 1996, 248 p.

Methodology: Generative IA for the Innovative Reuse of Reusable Materials

This work of transforming and reversing the architect's practices begins with identifying and rediscovering local materials and products available for construction. Often in the hands of builders or construction economists, the choice of materials, their specific characteristics, or their origin, in relation to their impact on the overall life cycle of the building, is once again becoming a fundamental element of the project for architects.

Deconstruction materials, industrial or artisanal production waste products, unsold or returned goods: the city, its businesses, and its users are full of potential resources that architects must explore in their production. The found object, repurposing, reuse, and recycling are strategies that contribute to the creativity and invention of the project, as well as to the development of an alternative materiality and image of the project. In response to a construction project, we, therefore, launched an experiment to discover this resource in a 1km radius around ENSAPVS as part of a pedagogical exercise in the first year of the Master's program at ENSAPVS.

In a second phase, the method developed in this research implements creative generative AI models, such as Text2Image, Midjourney, Dall.e, Gemini or Crayon, to propose and inspire a project design from a set of found and available materials for reuse. These generative models have been diverted from their use as simple graphic illustrations to develop design proposals with reusable materials and a specific architectural intention. These tools are interrogated by "prompts", a coding in natural language that makes it possible to describe and question a neural network trained on a set of graphic and syntactic data.

The third aspect of the research is the application of these techniques in a life-size construction site with a group of master's students from an architecture school. Here we take the example of the Archifolles pavilion, built by ENSAPVS students in May 2024, as part of a project funded by the Ministry of Culture to propose a cultural Olympiad to highlight architecture as part of the 2024 Olympic Games in Paris.

The Reuse of Materials in Architecture

The main disadvantages of reusing recycled materials that the students identified are as follows:

- Identifying and locating materials from construction sites, demolition sites, and industrial waste is difficult.
- Architectural design within the framework of a limited and heterogeneous stock of materials
- Development of scenarios for architectural details and structural assemblies

The proposed methodology is based on a process that uses the specific capacity of generative artificial intelligence text2image to count the number of material items and assembly suggestions according to specific criteria.

We analyzed five different GenAI packages to characterize their performance in counting and assembling a given set of materials. The use of generative AI on a limited and unpredictable list of material resources analyzes the quantity, shape, and physical characteristics that are disconnected from the preconceived architectural idea.

The experiment aims to evaluate the capabilities of digital design tools and generative AI to optimize the process of designing architecture with reusable materials. Therefore, it is essential to identify these capacities according to criteria inherent to reuse and construction in general. These could be materials, scales, quantities, dimensions, optimization and economy, assemblies and organization, as well as models and references.

2_Construcction process and installation of 2022 Pavilion by ENSAPVS students

Prompt Engineering for Architectural Ideation

To generate relevant proposals, text entries or "prompts" were created that focused on identifying clear variables highlighting the criteria relating to construction using reusable materials. In the experiments, the simplest possible prompts in English seemed to avoid possible biases by using references likely to lead to erroneous expected results.

We cannot detail all the generations produced on this project here, but here are four examples generated on two generative platforms, Midjourney and Dall. E.

On the following page (Fig. 3), the generated images present the material used in an innovative, economical, and sustainable approach. The variable "reuse" or "reclaimed" is then integrated into the "prompt" to estimate the capacity of the AIs to rethink the expression and implementation of the material. These images were generated by David Serero and Tim Houplain as part of the Data and Territories seminar at ENSAPVS.

By comparing these results, we can establish that the AIs seem to integrate the "reuse" variable well. Dall-E tends to create monumental projects, freeing itself from the economic criterion.

In the 2nd experiment, we tried to organize heterogeneous elements and provide innovative and original assembly solutions. (Fig. 4)

In this example, the appearance of the windows is heterogeneous. They are assembled in a vertical pattern with varying heights, but their widths are standard. The image proposed by Dall-E positions shutters perpendicular to the facade. An analogy can be made with the glass stiffeners sometimes used in the implementation of curtain walls, a term used in the prompt.

These two examples confirm that these generative AI platforms can propose arrangements of heterogeneous elements with an identifiable logic of assembly and structuring. The implementation proposed by these models raises questions that sometimes lead to irrelevant answers but are often able to inspire real design proposals. The students experimented with this method throughout their design process.

Critère	Prompt	IA génératives visuelles	
		Midjourney	Dall-E
Materialité	a contemporary architecture made of wood with an innovative, sustainable and economic design		
	a contemporary architecture made of reclaimed wood with an innovative, sustainable and economic design		

3_ Generative AI experiment of wood building in two differents model languages

Critère	Prompt	IA génératives visuelles	
		Midjourney	Dall-E
Assemblage	a facade made of windows of different sizes		
	a curtain wall made of windows of different sizes		
	a curtain wall made of reclaimed windows of different sizes		
	a contemporary curtain wall made of reclaimed windows of different sizes		

4_ Comparison of generation capacities on assembly windows with specific dimensions

Experiment 3 (Fig. 5) allows us to verify the ability of these models to quantify, enumerate, or give an order of magnitude of the quantities of materials or scales. The first language models of this type sometimes had difficulty identifying specific simple quantities (number of fingers or windows, for example). In queries made in the architectural context, the quantities are often large, and the scale is considered rather than the exact quantity.

Following numerous experiments of this type (Fig. 6), discussions with the pavilion's future users, and the project's technical constraints, the students were able to finalize a project. The article does not present the details of the stages of the hybrid design (classic design and augmented design) but some stages of the construction.

Here, reused materials were used for both primary and secondary structures. The case study of a few examples of reused materials today shows that reused materials are mainly used in secondary structures, as part of a "decorative facade" or as part of non-structural parts of a building, such as partition walls, doors, windows, etc. (Figs. 7 & 8).

Building a Scale 1 Architecture in Reused Wood

The final stage was the construction of the pavilion on a 1:1 scale on the site of the Olympic Village in the Parc de la Villette, which began with parts of the whole project using off-site construction up to on-site construction. This enabled the students to develop collaborative skills, enabling them to join forces and improve their intellectual and manual synchronization. An important part of this process was the immediate confrontation with reality. (Figs. 9 & 10).

Although we know that the construction experiment is only partial, given that its use has not yet been tested and its structure remains experimental, it has nevertheless demonstrated the ability to reinvent structural assemblies by reusing recycled materials.

5_ Midjourney result for a Prompt query: 50m2 box pavilion made of 2 million reclaimed wood lumber, in the style of Kengo Kuma

6_ Experimentation on a request for types of materials such as wooden Batten of the 70 mm x 70 mm type

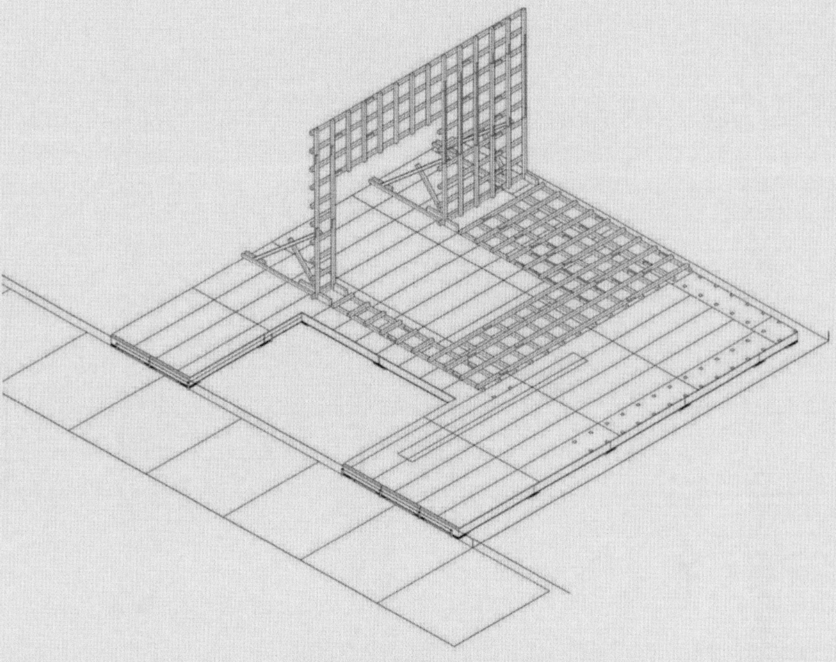

7_ Project "Le coup de poing." Construction method by folding the frame on the plane of the pavilion

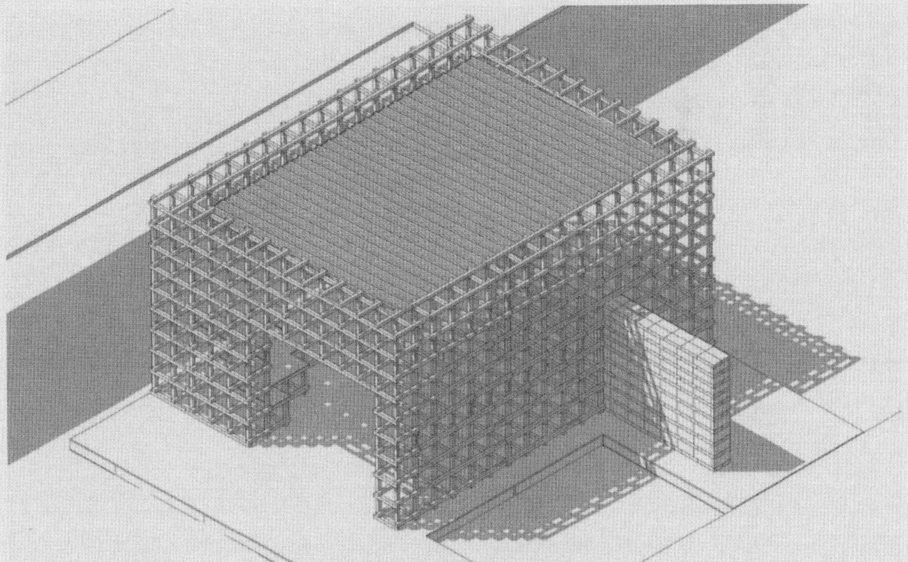

8_The pavilion is made up of 18 frames assembled by purlins in a three-dimensional lattice

9_Construction of the Pavillon on the site
of the Olympic Village in La Villette in May 2024

10_Detail of the reused rafter assemblies used in the pavilion

Towards a predictive Design of the Recycling Project

At the end of this research and the "Coup de poing" pavilion created for the Archifolies event, we identified that the urban environment is an important reservoir for recovering, safeguarding, storing, and reusing the materials and resources that makeup buildings and our living environment.

On the other hand, by having direct contact with the material and its constructive challenge, the architectural proposal emerges from the physical manipulation of materials, combined with generative AI experiments of the Midjourney and dall. E type, allowing for the expansion of creative experiments.

The design methodology, augmented by the help of generative design models and by interaction with one or more designers, makes it possible to envisage a predictive design model. This "cybernetic" collaboration portends interesting future creative models based on hybridization between generative models and our sensitive and cognitive skills. This evolution in the functioning of capacities is identified by a neurologist such as Raphaël Gaillard[5] as a new hybridization, such as that generated by reading and writing. It should make us reflect on an alternative pedagogical model to traditional architectural design learning.

5 Gaillard (Raphaël). *L'homme augmenté, Futurs de nos cerveaux.* Grasset, 2024, EAN : 9782246835172, 352 p.

Augmented Reality: Towards Inclusive Immersive Spaces

Video games, immersive spaces, and metaverses—these proliferating digital environments and the expanding array of associated services and activities are progressively increasing the time we spend immersed in virtual spaces. These environments now rival our primary reality, and at times, our perception blurs the distinction between them, making the virtual experience inseparable from the real. As we spend more time in digital immersive spaces than in physical reality, what will become of our relationship with our own bodies?

To explore this phenomenon, we must first examine the origins of immersive spaces and the challenges they present. We then propose an alternative approach to their design—one that does not conceive them as isolated fictional realms but rather as an **extension** of our primary reality, an augmented dimension that seamlessly integrates with our physical world.

This perspective allows us to envision identifiable **access points** between digital and physical spaces, fostering interaction and continuity between both realms. It has become crucial to integrate such portals into the design of immersive environments, ensuring they remain connected to our everyday spaces—whether private homes or public urban settings.

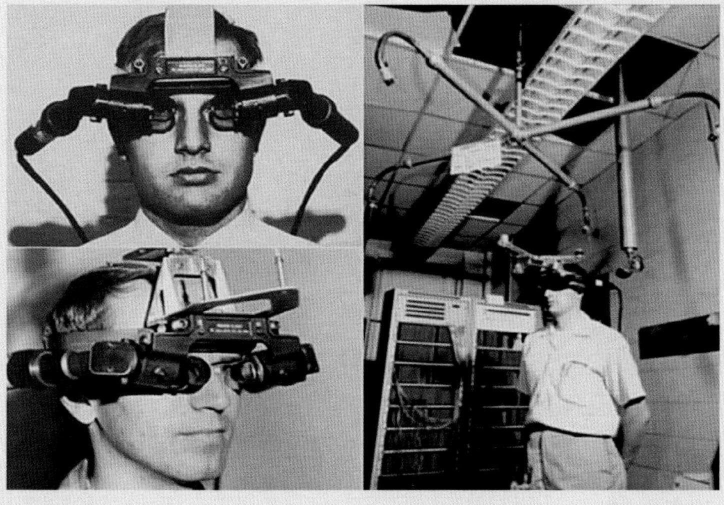

1_The Sword of Damocles (1968): World's first head-mounted display par Ivan Sutherland. Source: Devopedia.

The increasing fusion of digital technologies and services within urban environments is fundamentally reshaping the way cities are structured and experienced. From digital modeling of urban spaces to the dynamic processing of data that continuously redefines them, **mixed reality** introduces a novel form of interaction between the digital and physical worlds (Lucas, 2012). It enables geographically distant individuals to collaborate and collectively influence the construction of a shared reality. By drawing upon both **material and digital legacies**, mixed reality compels us to rethink hybridization processes, thereby transforming our perception of urban environments. In this redefined paradigm, the city regains its fundamental essence: a **space shaped by human interactions.**

1. The Disconnection Between Virtual and Physical Realities

Despite technological advancements, the vast majority of immersive environments remain **detached parallel worlds**, offering no tangible connections to the physical realm. When we don a virtual reality headset or another access device, we do not **transition** into an augmented reality—we instead experience an abrupt **disjunction**, stepping into an entirely separate dimension. There are no thresholds, no bridges, no fluid paths facilitating movement between these spaces. Instead, these environments often emulate video game landscapes or cinematic universes, devoid of spatial continuity with the real world.

For architects, this represents a profound risk—a future in which individuals lead fragmented lives, oscillating between an increasingly disconnected virtual existence and a dwindling engagement with physical space. Such a reality poses an existential challenge to architecture itself: if spatial design becomes purely digital, **what remains of the architect's role in shaping lived experience?**

2. Rethinking Immersive Environments: Augmenting instead of Replacing Reality

Our analysis and architectural expertise lead us to advocate for a different approach: the **design of augmented reality environments** that use physical space as a **framework** for virtual expansion. In this model, reality serves as an **anchor point**, allowing individuals to engage with virtual layers while remaining rooted in familiar spatial references. This strategy ensures that immersive experiences do not exist in isolation but instead integrate **seamlessly with the real world** (Huggett, 2020).

A persistent divide exists between **digitized heritage and virtual reality**, raising concerns regarding **presence and realism** in immersive environments. The contrast between **historical authenticity** and **digitally simulated spaces** creates a fundamental paradox. One proposed solution is **"free realism"**, an approach that seeks to bridge this gap by aligning digital reconstructions with lived experience. The challenge is to recognize virtual reality not merely as a simulation but as a **new paradigm of information systems and metaverses**—a framework through which digitized heritage can be fully integrated into historical exploration and interpretation.

However, emerging sociological and medical studies (Bojic, 2022) indicate that extended immersion in virtual environments is leading to significant **addictions and cognitive dissonance**, causing users to lose their spatial and bodily awareness. The metaverse, therefore, is not only a **platform for digital experiences** but also a **generator of spatial insecurity and psychological dependence.**

2_ Interior view of the " White house" for the metaverse by Andrés Reisinger and Alba de la Fuente

3. The Emergence of Imaginary and Parallel Worlds

The human ability to conceptualize and mentally inhabit **fictional, mystical, and imagined spaces** dates back to prehistoric times. The **cave paintings of Chauvet**, over 60,000 years old, illustrate early forms of storytelling and immersive visualization. The cave itself, isolated from the external world, functioned as a **portal**—a space where flickering torchlight animated images, transporting early humans into spiritual, historical, or aspirational realms.

Modern immersive environments, particularly metaverses, share striking similarities with these ancestral practices. Enabled by advances in **virtual reality headsets, high-speed networks, and real-time 3D rendering**, today's digital spaces offer **extraordinary levels of immersion**—often making the real world seem **dull by comparison.**

The term **metaverse**, first coined by Neal Stephenson in Snow Crash (1992), describes a **secondary reality**, freed from the constraints of physics and materiality. This world operates through **portals, teleportation hubs, VR devices, and even biochemical enhancements**—a spectrum of access points that blur the line between physical and digital. Emerging technologies such as **blockchain, NFTs, and digital twins** reinforce this trend, creating a virtual economy that exists parallel to the real world.

Fig 3_Travis Scott's unprecedented concert, which took place in April 2020 in Fortnite, attracted 28 million viewers at the height of the event, and in November 2020.

Yet, this immersion acts as a **double-edged sword.** It **liberates** the mind, yet **ensnares** it. Much like **sugar, tobacco, social media, and streaming platforms**, digital immersion fosters **dependency**, monopolizing cognitive and emotional space.

4. Architecture's Role in the Virtual Age

For architects, the metaverse presents both **a challenge and an opportunity.** The absence of **physical constraints**—gravity, material resistance, structural limitations—allows for **unprecedented creative freedom.** Since the 1990s, architects and digital artists have explored these possibilities, from **Greg Lynn's digital morphologies** to **Christopher Alexander's computational urban theories.** The emergence of paperless design studios at Columbia University and MIT's **Media Lab** further demonstrated architecture's role in shaping virtual environments.

However, architecture today risks **becoming irrelevant** within the metaverse. Unlike physical design, where spatial experience is **tangible and sensorial,** virtual environments often lack **a deeper architectural language.** Many game-like spaces—*Fortnite, Warfare*—are visually compelling but architecturally **superficial**, reflecting a culture more influenced by **graphic novels and cinema** than by **architectural thought.**

A key concern is **the environmental cost** of these digital realms. The real-time rendering of immersive 3D worlds demands **massive computing power**, contributing to **energy consumption on a global scale.** By 2024, data centers dedicated to immersive platforms are expected to account for **15% of global electricity use.** Should we not, therefore, **seek more sustainable alternatives**, where digital augmentation **complements rather than replaces** physical space?

5. Reclaiming Space: Towards a Hybrid Model

In response to these challenges, we propose an alternative vision: **the creation of hybrid, augmented reality environments.** These spaces integrate **physical and digital elements**, ensuring that immersive experiences remain **anchored to the real world** rather than supplanting it.

Through research and experimentation, we have developed **models of mixed reality** within the **Architecture & Intelligence Chair**

and the Virtual Reality Lab at Université Paris Cité. Inspired by **Peter Zumthor's atmospheric approach**, we emphasize **sensory engagement**, ensuring that immersive experiences **heighten rather than replace** spatial perception.

By establishing **real-world portals** into digital realms, we prevent the **total detachment** of individuals from their physical environments. We must **resist the dissolution of space** and advocate for a **future where architecture remains tangible, interactive, and human-centered.**

Should we surrender to **parallel realities**, or should we **invest in more meaningful, multisensory, and intelligent spatial experiences?** The answer will define **the future of architecture—and our place within it.**

Selected List of Projects

2021-2024 GARDEN OFFICE
Reclaimed Wood construction in a tree garden for a co-working and artisanal space.
Construction SERA, local reclaimed wood and roof tiles, and bricks.
On site fabrication and construction. Full BIM.

2018-2022 RESTRUCTURING OF THE GHANA AMBASSADY Paris 16th arrondissement.
Renovation of a private mansion. Complete project management. BIM Level 4

2015-2021 CLIMATE PLANET RESEARCH AND DATA CENTER
University of Grenoble Alpes, Grenoble. Competition winner. BIM Level 3, delivered in 2020.

2018-2021 RESTRUCTURING AND EXTENSION OF THE BOURGES IUT
University of Orléans. Complete project. Building and Furniture.
Creation of a cluster amphitheater for pedagogical innovation.
Competition winner, delivery 2021. BIM Level 3.

2016-2019 BAYEUX MEDIA LIBRARY CONSTRUCTION
Full project. Building and furnishings. Media library, auditorium, exhibition hall, Fablab workshop, conservation fund, Bayeux Intercom, (14)
Competition winner, 2019 completion. Full BIM. FACADE2BUILD 2020 award.

2016-2018 CIVIL SERVICE TRAINING CENTER (CNFPT)
Innovative training and conference center. Issy-les-Moulineaux
Competition winner, completed in 2018.

2013-2016 LEZOUX INTERMUNITIVE MEDIA LIBRARY
Full project. Building and furniture. Multipurpose room, Fab Lab, games library, and collections. Competition winner, completed in 2017, BIM-based DCE.

2011-2014 BATTLE OF FROMELLES MUSEUM
Museum and scenography. Competition winner, completed in 2014. DCE in BIM.

2010-2015	MEUDON-LA-FORET CULTURAL CENTER Performance hall, art workshops, and urban development or the neighborhood. Competition winner, completed in 2015.
2012- 2014	DANIEL ASSERAY CENTER Construction of a new social services center for the Nantes metropolitan area. Competition winner, completed in 2014.
2011-2014	SAINT-HILAIRE-DU-HARCOUET MEDIA LIBRARY Competition winner, completed in 2013. Named Best Public Building in 2014 by the Manche CAUE (French Community Economic Community).
2010-2013	AMIENS IUT LIBRARY AND AMPHITHEATRE Contest winner, completed in 2013.
2009-2013	CREATION OF THE BASEL/MULHOUSE AIRPORT BUSINESS CENTER Competition winner, completed in 2013.
2005 - 2007	AFRICAN DIASPORA AND BENIN ART EXHIBITIONS, MUSEE DU QUAI BRANLY, Paris Curator: Claire Denis. Competition Winner, opening: October 2.
2004-2005	VARIABLE GEOMETRY ACOUSTIC DOMES Installation of a variable-shape acoustic shell for the performance hall of the Villa Medici in Rome.
2004-2010	HELLENIKON PARK AND URBAN DEVELOPMENT, Athens, Greece Winner of the international competition for the redevelopment of the former Athens International Airport site, covering 400 hectares of parkland and 130 hectares of urban development. Phases completed – preliminary design, preliminary planning, and project development – then abandoned due to the Greek government's bankruptcy.

Architecture Awards

2021 FACADE 2 BUILD AWARDS WINNER
Bayeux Media Center – Normandy

2020 GUSTAVE EIFFEL PHD AWARD
Architecture & energy research, GSA Lab – Paris

2016 INNOVATION NATIONAL AWARD
Lezoux Media library – Grand Prix Livres Hebdo des bibliothèques francophones

2014 BEST PUBLIC BUILDING AWARD
Saint-Hilaire-du-Harcouët Media library

2007 FIRST PRIZE WINNER INTERNATIONAL COMPETITION FOR METROPOLITAN PARK, Greek Ministry of Environment, Urban Planning and Transportation, Athens

2005 FRENCH ACADEMY IN ROME
Rome prize winner at Villa Medici for variable geometry acoustical domes

2004 NEXT GENERATION PRIZE WINNER
Scale Orography Open air theater project, Metropolis Magazine, New York

2004 VIII VENICE BIENNALE
Architecture exhibition at the French Pavilion in Venice

2003 YOUNG ARCHITECT AWARD
Best Architectural Practice , New York's Architectural League

2002 NOUVEAUX ALBUMS DES JEUNES ARCHITECTES
Winner of National Young Architect award at DZo Architects (Descombes, Fernandez, Regnault & Serero)

1997 FULBRIGHT SCHOLARSHIP FOR COLUMBIA UNIVERSITY
French-American Foundation, Paris

About the Author

David Serero is architect and urbanist, known for his pioneering research in material innovation, geometry, and sustainable architectural design. Born in Grenoble, France, he is a graduate of Columbia University in New York and the École d'Architecture Paris-Villemin. In 2004, he was awarded the prestigious Prix de Rome, an honor that significantly shaped his professional trajectory. After several years in New York, he returned to France, establishing his practice in Paris in 2006.

Serero's work is characterized by a distinctive exploration of materials and structural geometries, with a particular focus on architectural skins. His design for the Bayeux Media Library exemplifies this approach, integrating a porous structural system that optimizes natural light diffusion for reading spaces and multimedia experiences.

His architectural projects have been widely exhibited both in France and internationally, featuring in esteemed institutions such as the Frac d'Orléans, the Pavillon de l'Arsenal, MoMA in New York, the Architectural League of New York, Villa Medici in Rome, and the Mori Museum in Tokyo.

In parallel with his architectural practice, David Serero is an associate professor at the École d'Architecture Paris-Val de Seine, where he teaches and conducts research. His PhD (2020) explores the untapped potential of urban environments for wind energy harvesting, leveraging building edges and façade surfaces as active energy-generating elements.

Serero is also the co-founder and co-director of the Research Chair Architecture & Intelligences, as well as co-director of EVCAU, a research laboratory at Université Paris Cité, which investigates how material resources and digital data are transforming architectural practice. Throughout his career, he has held teaching positions at Pratt Institute in New York, Columbia University, Paris Sciences et Lettres University, and Université Paris Cité, while leading numerous architecture and research workshops across France, Italy, Greece, Spain, Austria, and the United States.

Contributors and Collaborators

We would like to express our heartfelt gratitude to all those who have contributed to the work of the Office over the years.

Associates • Iñaki Abalos, Gabriela Barman, Philipp Barman, John Beckmann, Amir Ben Hassine, Anthony Chen, Philippe Coignet, Stella Daouti, Arnaud Descombes, Michael Drevet, Eric Dufour, Bernard Giraud, Houssem Jendoubi, Bruno Lavaux, Pascale Mathiault, Stéphane Maupin, Philipp Mohr, Ali-Reza Razavi, Antoine Regnault, Renata Sentkiewicz, Benjamin Serero, David Serero, Mike Silver, Youri Sobotko, Yves Ubelmann, Philipp Von Dalwig, Isabelle Yafil, Stéphane Maupin **• Research associates •** Edith Akiki, Gilles Bouchet, Philippe Bach, Loïc Couton, Vincent Dubois, Anne-Laure Herry, Florian Hertweck, Remy Kervadec, Robert Leroy, Alexis Meier, Jean Denis Parisse, Valère Paupelin-Huchard, Mario Poirier, Michael Silver, Dorothée Sipp, Junh-Ha Suh, Pascal Terracol, Dimitri Toubanos, Joel Sakarovitch, Hani Rashid, Lise-Anne Couture, Evan Douglis, Maurice Brocato, Christian Girard, Jean-Louis Cohen, Bernard Cache, Kenneth Frampton, Laurent Lescop, Bernard Tschumi, Mark Wigley, Yehuda Safran, Pierre Vincent, Jean-Marc Weil **• Collaborators •** Nadia Abourrif, Arrate Arizaga, Camille Ballon, Valentine Barouin, Méolodie Barreau, François Bay, Xavier Begel, Luca Bernardi, Romain Boboe, Corentin Boiteau, Pierre Antoine Bonamour, Anthony Boueiz, Marine Bouhin, Antoine Bourdeau, Julien Braga, Mathieu Brassart, Tristan Cadorel, Bastien Casasoprana, Mariella Chavez, Nordine Chevallier, David Chicheportiche, Mélanie Claude, Laëticia Cluzel, Charlotte Cochet, Glenn Cohen, Julien Defontaines, Charles Delcourt, Marie Fleur Delpy, Mahdi Derakshan, Louis Destombes, Ilhame El Amraoui, Anthony Fauve, Elena Fernandez, Jonas Fernandez Perez, Anna Fojcik, Francesca Fontana, Maxence Froment Meurice, Julia Garcia Diez, Rodrigo Garcia Sayan, Maria Garrido, Alice Godefroid, Anna Godefroy, Louisa Gouesnard, Paula Granadillo, Marta Guedan Vidal, Claire Guinguene, Mautrin Hachard, Noriko Harada, Mark Havasi, Kien Hoang, Maëlle Josselin, Audrey Kac, Julien Kemler, Nathael Ken Hassan, Hanife Koc, Julien Lanoo, Thuy Lant Tran Thien, Brice Lajeunesse, Solène Lavel, Fanny Lenoble, Lionel Leotardi, Katarzyna Lesniewska, Antoine Loyer, Dusanka Malicevic, Pascal Mendez, Louise Michaud, Uy Miennfa, Hui Min, Lloris Mkhoul, Perrine Monfort, Olivier Muller, Dominique Navarro, Charlotte Nocchi-Courthieu, Sabrina Noiraud,

Cecile Ortolo, Emily Ottinger, Olivier Oudin, Yoichi Ozawa, Amandine Parisot, Pierre Pasquini, Alexis Payen, Mathieu Planquette, Claude Polycarpe, Amandine Quillent, Lucie Reau, Andrija Rusan, Adrianna Rybkowska, Alice Sabatier, Philippe Sanfourche, David Serel, Marc Serena Montanana, Louise Serero, Ran She, Min-Seung Shin, Ludovic Smagghe, Jean Baptiste Simon, Grégoire Simonnin, Esther Sperber, Taichi Sunayama, Sylvain Thery, Elena Tira, Nemanja Topic, Clara Toubiana, Camille Verdu, Ondine Wartelle-Collard, Jaime Yndurain, Marie Pierre Yon, Hsin Yu Chen, Fabrice Zaini • **Artists** • Maurice Benayoun, Claire Denis, Yousri Nsrallah, Erwin Redl, Mierle Ukeles • **Photographs** • Didier Boy de la Tour, Stéphane Chalmeau, Billyana Dimitrova, David Henry, Arnaud Lesage, Olivier Tournon • **Curators** • Anne-José Arlot, Marie-Ange Brayer, Anne Behr, Marco Brizzi, Luigi Centola, Baptiste Delamare, Grégory Domine, Olivia Du Mesnil, Laure Groznykh, Sarah Herda, Aldo Micillo, Clifford Pearson, Richard Peduzzi, Luigi Prestinenza Puglisi, Anne Rieselbach, Béatrice Simonot, Marcello Smarelli, Konstantin Tsakonas, Abigaël Verrier • Contractors • Alliou Dlallo, Mouhieb Kirala, Henrik Kogut, Desiré Lagrenne Thierry & Ignacio Martins • **Clients** • Bonaventure Adjavor, Betrand Alard, Alain Badiou, Robert Baldwin, Anna Bossman, Jacques Bouille, Shawn Burns, Hélène Cerutti, Pedro Coello Behr, Jean-Pierre Cornet, Hilaria Dey, Paul Echart, François-Xavier Horvat, Jean-Claude Horvat, Jean-Christophe Lacas, Marc Lecorre, Mercedes Lecorre, Philippe Moysan, Shalom Nachshon, Jean-Marc Poirier, Dror Segal, Vanessa Sitbon, Ville de Bayeux, Ville de Lezoux, Ville de Saint -Hilaire, Hélène Tonnelet.

And all students and researchers that participated in the work and research from ENSA Malaquais, ENSA Paris Val de Seine, Avery Hall Columbia, Ecole Speciale Architecture, Pratt Institute, EVCAU Research Lab, GSA Research Lab, les Grands Ateliers Isle d'Abeau (GAIA).

I would like to thank my family, and especially my children, Emma and Alvaro, for their patience and limitless love throughout the many moments all this work kept me away from them.

French Translation

Avant-propos
DONNÉES | ÉNERGIE | MATIÈRE ARCHITECTURES CLIMATIQUES

David Serero

Cette exploration de vingt ans de pratique architecturale est un temps suffisamment long pour assister à la sédimentation d'idées et d'approches fondamentales dans la conception architecturale. Les projets sélectionnés ici révèlent une récurrence de figures et de concepts clés qui se déploient et se dissipent tout au long du travail. Ils remettent en question la notion d'architecture en tant que structure permanente, préconisant plutôt une approche transitoire et éphémère avec une plus grande polyvalence spatiale. À l'image d'un réseau de mycélium. C'est une "architecture ouverte", qui favorise le partage, la création de communautés et l'interconnexion.

Notre observation approfondie des processus naturels nous a appris que le temps est le meilleur allié de l'adaptation, de la transformation et de la co-création. Les arbres, souvent utilisés comme du mode dans notre travail, incarnent l'optimisation et la redondance nécessaires à l'adaptabilité et à la transformation contextuelle, a influence d'architectes comme Alvar Aalto et Frank Lloyd Wright souligne la manière dont la nature inspire les projets et les expériences architecturales.

Au fil des années, nous avons appris que «construire» est intrinsèquement lié au concept d'énergie, non seulement en termes de consommation, mais aussi de production et de partage. Ce paradigme énergétique va redéfinir l'architecture. Les bâtiments de demain ne se contenteront pas d'économiser de l'énergie, ils en produiront et la partageront.

Nous voyons les bâtiments sous l'angle de l'énergie et de la résilience. Pour nous, l'architecture doit minimiser son impact sur l'environnement et s'intégrer à la nature. Comme un arbre, l'architecture doit se déployer dans son contexte, respirer avec son environnement et tisser de nouveaux liens avec son entourage. Notre réflexion s'inspire de modèles trouvés dans la nature, Elle ne vise pas à créer de nouveaux espaces, mais à déployer des modèles d'origines culturelles, sociales et spatiales.

Trois sources centrales émergent des projets présentés dans cet ouvrage. Plutôt que les trois principes de l'architecture, ses sources sont plutôt à considérer comme les éléments déclencheurs de l'architecture plutôt qu'un objectif de l'architecture.

Donnés

Les données numériques et les codes de programmation constituent un nouveau champ de recherche que les architectes doivent acquérir et mobiliser ces connaissances dans leur travail. L'architecte-chercheur met ainsi en place des dispositifs de génération et d'analyse de données, et développe un mode de conception "informé" de type Design by Data. La possibilité pour l'architecte d'interagir directement avec les outils d'analyse lui permet de mieux comprendre l'impact de chacun de ses choix sur l'environnement du projet, son contexte urbain et territorial. Cette méthode de conception architecturale complète les méthodes utilisées par les architectes: conception parallèle d'une échelle à l'autre du projet, affinement du sens global du projet jusqu'à son expression la plus fine dans les détails architecturaux, réalisation de prototypes et préfiguration de l'assemblage des matériaux du projet, qui sont des méthodes éprouvées pour analyser, comprendre et concevoir un espace.

La conception paramétrique, les flux de travail collaboratifs, la conception augmentée et les processus pilotés par l'IA ont transformé nos projets. Les mathématiques de l'architecture ont évolué, passant des dimensions et des proportions à des processus techniques pilotés par des écosystèmes de données. Ces avancées nourrissent le processus de conception, mais elles remettent également en question le rôle traditionnel de l'architecte. Comprendre la logique des machines, des réseaux neuronaux et des algorithmes sera essentiel pour envisager de nouveaux modes de collaboration entre l'homme et la machine.

Dans cette nouvelle ère des données, les architectes risquent de perdre leur leadership dans la création architecturale, car son processus est de plus en plus distribué entre les personnes et les outils. L'avènement de l'IA générant des textes-images, des textes-modèles, et bientôt des discours-architecture, remet en question l'un des derniers bastions du rôle de l'architecte: l'idéation. Les architectes deviendront-ils de simples assistants des machines? Nous croyons fermement que la maîtrise de la logique des machines, des réseaux neuronaux et des algorithmes est cruciale pour contrôler et améliorer la collaboration entre la créativité humaine et l'innovation technologique.

Energie
Notre philosophie définit la question énergétique comme indissociable de la question architecturale. Energie et architecture sont intrinsèquement liées, formant un objectif commun qui façonne à la fois le design architectural du projet et son approche environnementale.

L'architecture est aussi un environnement dynamique qui génère des ambiances intérieures, régule le rayonnement solaire et la ventilation ventilation naturelle et limite les besoins énergétiques. Presque tous nos projets intègrent des dispositifs bioclimatiques en tant qu'éléments fondamentaux, façonnant l'image même de nos conceptions.

Notre recherche s'est largement concentrée sur l'intelligence et l'adaptabilité de l'enveloppe de bâtiments. Récemment, dans le cadre d'une initiative de recherche globale, nous avons étudié la ventilation naturelle et la circulation de l'air autour des bâtiments. En nous appuyant sur des logiciels de simulation avancés, nous avons développé des méthodes pour faciliter les décisions de conception à l'aide de dendrogrammes arborescents, permettant des expériences virtuelles avec la lumière, le comportement du vent et le rayonnement thermique.

Elle anticipe une ère de symbiose énergétique, où les bâtiments ne sont pas de simples consommateurs mais des nœuds indispensable dans un réseau de partage de l'énergie. La micro-production d'énergie est devenue de plus en plus importante, les dispositifs étant progressivement intégrés à la peau des bâtiments. Cette évolution est favorisée par l'émergence de réseaux intelligents ou de l'internet de l'énergie, où les bâtiments, les quartiers et les villes partagent et échangent de l'énergie, transformant chaque entité en un producteur et un distributeur d'énergie actifs.

Notre vision comprend des dispositifs d'énergie éolienne et solaire entièrement intégrés, combinés à des façades adaptatives qui s'ajustent passivement et continuellement aux conditions locales. Cette approche transformatrice garantit que l'architecture n'est pas seulement réactive, mais qu'elle participe activement à la création d'environnements durables et libérés des contraintes énergétiques.

Matière
La notion de matière est une composante majeure de la conception du projet selon trois axes: Matière-histoire, Matière-expérience ou Matière-ressources. Elle est déployée, pour créer un dialogue avec l'histoire du lieu dans certains projets, ou pour créer une expérience physique ou conceptuelle de la matière et dans d'autres, ou encore comme mode constructif alternatif en matériau de remploi dans une pos-

ture militante de préservation des ressources naturelles.

Dans le contexte de villes anciennes et de bâtiments patrimoniaux, la matérialité est une force de dialogue permettant de tisser un lien à travers les âges entre architecture ancienne et architecture contemporaine.

L'expérience de la matière dans le projet place le matériau avant l'espace ou le volume. C'est une force génératrice de l'architecture.

Avec la matière-ressource, nous enracinons profondément nos structures dans leur environnement, en utilisant des matériaux locaux pour activer un cycle de matériaux durables avec des ressources récupérées, trouvées ou mises au rebut.

Cette approche favorise un changement radical: les bâtiments ne sont plus de simples constructions mais des banques-ressources de matériaux, faisant écho aux pratiques de nos prédécesseurs qui réutilisaient la pierre et le bois des structures existantes. Avec la raréfaction des matériaux de construction et notre engagement inébranlable à réduire l'empreinte carbone des bâtiments, nous sommes confrontés à une vérité vertigineuse: les architectes devront construire moins et probablement se concentrer sur ré-imaginer des structures existantes ou des sites déjà urbanisés, au lieu d'en créer de nouvelles.

Pour nous, l'architecture est une "matière informée". Elle est le catalyseur du cycle émergent des matériaux. Elle incarne la révolution numérique, révolutionnant le travail créatif dans toutes les dimensions du projet.

Cette analyse de notre pratique de l'architecture guidée par les données, l'énergie et la matière. Ces projets combinant méthode expérimentale, participent ainsi à la définition de ce qu'est la recherche en architecture. Si l'architecte développe, dans certains cas, de nouvelles connaissances sur le projet, sa conception ou sa construction, ne génère-t-il pas une approche qui constitue une contribution à l'épistémologie de l'architecture ou aux méthodes de conception?

---

1. DATA

Des processus naturels aux réseaux neuronaux artificiels

L'usage des données dans la conception du projet ouvre une nouvelle relation entre mathématiques et modèle de conception du projet. Si la discussion sur les dimensions et les proportions du bâtiment était centrale dans les anciens traités d'architecture, l'avènement de l'espace cartésien, du modèle numérique, de la conception paramétrique, de la conception augmentée par des grands modèles de langage d'intelligence artificielle ont ouverts de nouveaux potentiels conceptuels et techniques pour l'architecture.

Si ces avancées nourrissent et irriguent le processus de conception de nouvelles méthodes, elles remettent également en question le rôle traditionnel de l'architecte. Elles incitent à repenser sa mission en intensifiant la collaboration entre l'homme et la machine basé sur la maîtrise de la logique des machines, des réseaux neuronaux et des algorithmes. Les projets présentés ici adressent un certain nombre de réflexions sur ces enjeux.

Le projet GhostTrack pour la biennale de Venise de 2002, réalisé avec A. Descombes et A. Regnault au sein du groupe DZo, posait de manière prémonitoire, la résistance et l'incapacité de modèles numériques et de technologie 3D à réaliser des opérations fines sur la matière, telles que les principes de tressage des matériaux pour former une structure. L'installation Ghostrack au pavillon français, était conçue sur la base d'un modèle numérique et réalisée à la main, à l'inverse d'autres œuvres de cette époque, basées sur une fabrication grâce à des technologies d'impression 3D de géométries conçues de manière classique.

Les projets de l'auditorium de Saint Cyprien, le dôme à géométrie variable à Rome, le centre culturel de Meudon-la-Forêt et l'ADN Eiffel sont des explorations sur les technologies génératives de géométrie pour le projet. Un modèle paramétrique est codé puis utilisé comme processus de génération de formes pour le projet qui irriguent toutes les échelles de la conception. Dans ce modèle, ces codes s'inspirent souvent de processus génératifs naturels—eux-même issus de logiques algorithmiques (Wolfram, 2002)—qui sont détournés ou appropriés dans le cadre du projet.

Pour l'auditorium de Saint Cyprien, des modèles de branchages et de port d'arbres sont codés avec l'objectif de générer un complexe de toiture et de plafond ajourés pour créer sous-bois lumineux qui englobe tous les espaces du bâtiment. Un modèle similaire a ensuite été codé pour générer le feuillage paramétrique pour la forme des baies oblong du centre culturel de Meudon-la-Forêt.

D'autre part, le projet ADN Eiffel a complètement numérisé le profil structurel de la tour Eiffel dans un modèle paramétrique pour créer par prolongement de ce profil, une extension de la forme, tel un bourgeonnement de cette structure centenaire, afin de proposer une plateforme provisoire pour la tour Eiffel.

Dans le travail réalisé dans le cadre de notre résidence à la Villa Medici à Rome, le projet de dôme à géométrie variable est issu d'une recherche sur le déplacement d'ondes acoustiques dans les modénatures des espaces intérieurs classiques d'une grande salle de réception. Ce modèle de propagation a permis de définir le profil d'une surface variable de coque acoustique permettant d'adapter la réponse acoustique de l'espace aux types de performances: chant, musique de chambre, musique amplifiées, conférence. Autant de scénarios d'usages où l'adaptabilité de la forme permet une modification de l'expérience acoustique d'un événement.

Dans le cas du projet Modulations pour un jardin temporaire au Québec, c'est le modèle de la croissance végétale de la fougère qui a été réutilisé pour créer un écran d'acier inox à la fois jumeau végétal et guide sinueux des visiteurs du jardin.

Pour le learning Center à Issy-les-Moulineaux, la restructuration du bâtiment prend un modèle fractal, celui de l'éponge de Menger-Sierpinski, comme stratégie de reconfiguration et d'évidement progressif des espaces intérieurs du bâtiment afin d'allouer à des espaces classiques de formation des espaces informels et libres (Schugurensky, 2007). Cette stratégie permet de créer des espaces pédagogiques hybrides, où le temps passé dans et en dehors de la classe participe au renforcement des apprentissages.

Architecture et données

L'organisation de l'espace et du temps a longtemps été le privilège de l'architecte. Le projet architectural est autant une organisation des espaces, des usages ou des styles, qu'une projection dans le temps futur d'un lieu. Penser le futur à partir des traces du passé met l'architecte dans une situation d'archiviste du possible.

Mais la discipline architecturale a été utilisée comme modèle à plusieurs reprises pour développer de nouvelles méthodes d'organisation ou de structuration des connaissances, comme l'illustrent les deux exemples suivants: les méthodes de mémorisation ou la programmation informatique ou les langages de codage.

Par exemple, depuis l'Antiquité, l'expérience des lieux architecturaux a été un puissant support pour notre mémoire. Comme le raconte Francis Yates dans L'Art de la mémoire, dans la Grèce antique, des techniques de mémorisation appelées "palais de la mémoire" ont été développées pour entraîner la mémoire des hommes de science. Dans ce livre, Yates raconte que l'espace physique est la base d'un espace mental ou fictif utilisé comme mode de mémorisation des données. Cette "méthode des lieux" anticipe, d'une certaine manière, l'émergence de l'hypertexte (Bush, 1945) , un nouveau mode de lecture et de partage des connaissances par la diffusion de textes non linéaires , organisés par affinités sémantiques ou thématiques.

Un autre exemple de la relation étroite entre l'architecture et le modèle organisationnel se trouve dans le livre A pattern language. Dans cet ouvrage, l'architecte anglais Christopher Alexander analyse en 1977 l'organisation des villes et leurs composantes: les espaces urbains. La ville, en raison de sa grande échelle, peut être décomposée en micro-espaces publics, de plus petite échelle, de nature élémentaire qui pourraient constituer, à partir d'un nombre limité de configurations, la matrice de toutes les villes connues. Selon Alexander, un langage de 253 "motifs premiers" refait surface dans toutes les villes connues. Cette pensée, étonnamment, aura plus tard un impact dans le domaine de l'informatique, notamment dans la création de langages d'objets, que dans l'architecture. En identifiant des similitudes (motifs) dans l'organisation complexe des villes et en démontrant leur récurrence, cette recherche a eu un impact considérable sur l'émergence des langages de programmation informatique et l'organisation des données numériques.

Ces protocoles ont progressivement permis la mise en place d'une syntaxe commune à la mondialisation—la mise en relation des économies mondiales—depuis les années 1990. Ils ont également participé au déploiement des procédures d'agrégation et d'interrogation automatiques des données, puis, dans une période plus récente, à l'émergence de nouveaux modes de collaboration (par exemple, la conception continue pour l'industrie) et d'échange social (réseaux sociaux numériques) d'une ampleur et d'une portée jamais envisagées auparavant. Cela a conduit plus tard au développement de plateformes numériques multiservices mondiales, réunissant réseaux sociaux, services bancaires, services de mobilité et plateformes de travail collaboratif.

Les technologies numériques se sont ainsi infiltrées dans nos activités sociales et culturelles et ont profondément modifié la pratique de notre espace territorial et de notre habitat. Nos activités physiques et virtuelles ont généré une multitude de traces numériques que les systèmes informatiques ont consolidées en lacs de données. Cette masse de données (Big Data) associée aux développements de l'apprentissage automatique (deep learning) ont permis le développement d'algorithmes qui, par amélioration itérative, sont capables d'identifier des éléments aussi complexes que le niveau de sécheresse des champs agricoles sur des photos satellites territoriales, des embryons de tumeurs sur des images médicales, ou de prédire nos désirs à partir de nos pratiques de consommation en ligne. Plus le jeu de données est important, plus la précision des prédictions est fiable, comme le montrent les courbes d'apprentissage de l'évaluation de la calibration de ces systèmes. Dans cet exemple, en multipliant le champ de données par 3 (de 1000 à 3000 exemples), ce système augmente la fiabilité des résultats de 20 % (de 70 à 90 % de performance) et dépasse la performance de la classification humaine. Cette approche mathématique, maladroitement appelée intelligence artificielle (IA), a poursuivi la révolution industrielle avec l'automatisation d'un nombre croissant de nos activités, cette fois non pas physiques (révolution industrielle), mais surtout intellectuelles (systèmes experts), avec des niveaux de précision et de prédiction que nous n'avions pas anticipés.

L'impact conceptuel de ces technologies sur l'architecture reste largement sous-estimé aujourd'hui, malgré les travaux de recherche qui ont investi le domaine de la conception architecturale ou urbaine, tels que la capacité des machines à penser (Turing, 1951), la conception architecturale autonome—"Architecture-By-Yourself" (Negroponte & Friedman, 1976), la programmation de réseaux de neurones pour l'apprentissage (LeCun, 1991), la personnalisation de masse dans un processus de fabrication numérique (Cache, 1997), ou la recherche des limites de la technologie de l'IA créativité et empathie (Kai Fu Lee, 2018), et la conception automatique de projets par le langage naturel (Dall.E, Open AI, 2022).

Dans les phases de réflexion et de production du projet architectural, ces technologies aident, remplacent et déplacent le rôle des architectes à plusieurs étapes du projet. Ces évolutions participent à la transformation des missions des architectes et de leur travail dans une mesure qui reste difficile à quantifier, mais dont l'impact est certain et important à court terme. En réaction à l'émergence de ces technologies, certains architectes préfèrent négliger cette évolution et s'enfermer dans une pratique architecturale de type artistique (corbuséenne, néo-moderne ou poétique), en résistance à cette vague, plutôt que de penser le rôle de l'architecte comme garant de la qualité du bâti en collaborant avec ces systèmes.

D'autres cherchent des moyens de limiter et de restreindre la concurrence avec ces systèmes dans la complexité et les incohérences des réglementations qui monopolisent la profession. Mais les réglementations peuvent-elles rester longtemps en avance sur l'évolution rapide de ces technologies? Mon enquête sur ce sujet montre plutôt que c'est dans une méta-collaboration homme/machine que les architectes peuvent progresser et accroître leur créativité, leur productivité et faciliter la difficile coordination des multiples acteurs de la conception architecturale et de l'acte de construire.

Les architectes seraient ainsi au centre d'un écosystème de données sur la conception et la gestion des bâtiments et des villes, dont l'importance est sous-estimée et sur laquelle les compétences des architectes sont très partielles, voire inexistantes. La complexité de la construction des bâtiments, au sein d'équipes parfois très importantes, de la conception, de la construction et de la gestion de la vie du bâtiment, renforce l'importance du domaine des données riches et complexes qui reste à structurer et à consolider (BOS, 2021). Dans les années 2000, des industries telles que l'automobile ou l'aéronautique, grâce à de puissants opérateurs sectoriels (Dassault Systèmes, Stellantis, Airbus, etc.) ont réalisé une intégration complète du processus de conception en mode continu et partagé, améliorant la qualité des produits et réduisant les délais et les coûts de fabrication.

Aujourd'hui, la maquette numérique 3D du projet architectural, souvent appelée à tort BIM (Building information model), est plutôt un simple modèle de collaboration, même si la culture de l'ingénieur (rationaliser et optimiser) est rarement en phase avec celle de l'architecte (la cohérence artistique de l'ouvrage). Elle nous permet d'articuler la conception du projet par la modélisation intuitive, sa production architecturale par le cahier des charges intelligent, et de consolider le champ de données du projet pendant la construction. Les données numériques ainsi générées par le processus de construction ou par un bâtiment connecté sont une nouvelle source de données pour une épistémologie de l'architecture.

Vers une conception architecturale augmentée

Quel est le rapport entre la conception architecturale et les données numériques? Il nous semble que deux éléments nous permettent d'explorer cette question.

Comme évoqué plus haut, la production de données liées à nos usages depuis le début des années 2000, par les plateformes sociales ou commerciales, par les appareils numériques mobiles qui nous suivent partout, mais aussi plus généralement par l'Internet des objets, constituent d'immenses lacs de données (appelés Big Data) que nous avons archivés et rendus accessibles. Ces sources, aux structures variées, constituent un excellent gisement d'exemples, de comportements, d'actions/réactions que nous pouvons interroger pour prédire des comportements ou répondre à un certain nombre de requêtes.

Le second aspect est la maturité récente de l'apprentissage profond, qui associé à plusieurs techniques de programmation de réseaux de neurones permet de répondre de manière de plus en plus précise à des objectifs de prédiction, de requêtes d'analyse ou de modélisation générative d'images ou de modèles 3D. Nous pouvons citer ici deux exemples:

–Les applications récentes de Text2image en ligne ont atteint une efficacité et une qualité, comme le démontre la série d'applications telles que MidJourney, Dall.E ou Stable Diffusion. Basés sur une description sémantique, ces outils imaginent, dessinent ou génèrent une image basée sur l'assemblage de références visuelles correspondant à la requête de manière quasi immédiate.
–ChatGPT, nouvel agent conversationnel, obtient également des résultats impressionnants en termes de fiabilité, de qualité et d'exhaustivité des réponses qu'il génère sur des sujets aussi vastes que la philosophie, la technologie, l'éthique et les mathématiques. Comme le mentionne Yann Lecun dans 2022, ces systèmes progressent mais restent dépourvus d'une conscience commune, permettant d'arbitrer ou de guider des choix ou des actions.

L'impact de cette technologie sur la réflexion architecturale est multiple: le déploiement du DBD (design by data), l'idéation automatique à partir de briefs clients, la génération de plans ou de maquettes 3D à partir de consignes, l'optimisation de systèmes à paramètres multiples comme le coût de construction, ou encore l'optimisation de la gestion énergétique ou de la ventilation.

Mais l'économie des données, dans laquelle nous sommes entrés, n'a pas un impact environnemental nul. Nous pouvons mentionner, entre autres, l'importante énergie consommée pour rafraîchir les processeurs des gigantesques centres de données qui stockent nos données, mais aussi nos logiciels. Mais les processeurs à haut rendement énergétique s'améliorent rapidement et surtout l'émergence de l'informatique quantique, nous permet d'envisager une multiplication de la puissance de calcul tout en conduisant à une baisse inversement proportionnelle de l'énergie consommée.

Les données et en particulier le suivi et la surveillance des données environnementales sont le facteur clé pour détecter les changements dans notre environnement tels que: la mesure des températures, la mesure des vents à la surface de la terre, les mesures de la composition chimique de notre atmosphère, le niveau de glace et de neige sur la terre, le recul des zones forestières et l'augmentation des zones urbaines et du trafic humain.
Ces données ont révélé la disparition rapide du nombre d'espèces de la biodiversité et nous permettent de comprendre l'évolution de ces

phénomènes qui, pendant des milliers d'années, n'étaient pas visibles à l'échelle d'une génération humaine.

LE TEXTILE DANS L'ARCHITECTURE ET L'UTILISATION DE L'ORNEMENT PAR RAPPORT À LA STRUCTURE

Ne pourrait-on pas dire que toute forme inventée se manifeste comme le pli – jusqu'au froissement, jusqu'à l'informe et la délacération – d'une forme précédente? C'est une forme de drapé dans tous les cas.
Georges Didi-Huberman

1. Dans son entreprise de réhabilitation du concept warburgien de *survivance* (*Nachleben*), est-ce un hasard si Georges Didi-Huberman choisit précisément le drapé tombé pour le projeter dans un *modèle fantomal* de l'Histoire de l'Art[1]? La *survivance* suppose toujours un mouvement de périphérisation, qui procède à la fois d'une évaporation et d'un dépôt, d'un effacement et d'un archivage. Une mémoire au travail qui contrarie la continuité de l'Histoire des Formes. La structure d'exclusion c'est ici le mouvement de l'étoffe/voile/linceul, qui offre tout à la fois la nudité du corps de la Nymphe au regard, et son autonomie formelle au drapé. Au bas du tableau, à sa périphérie et presque déjà en dehors, le drapé/plissé/tombé cultive sa spectralité et sa *puissance résiduelle*. Derrida écrit:

Toujours une forme sur un fond mais le *parergon* est une forme qui a pour détermination traditionnelle non pas de se détacher mais de disparaître, de s'enfoncer, de s'effacer, de se fondre au moment où il déploie sa plus grande énergie.[2]

C'est déjà tout le paradoxe du *parergon*, sa localisation ambiguë, qui est posé: l'*en dedans* de l'œuvre comme organiquement lié à son *en dehors*, en ce qui en trace la limite en la recourbant et la dispersant vers l'intérieur. Comme Aby Warburg avant lui, Didi-Huberman nous invite à considérer la permanence des formes, leur *survivance*, non pas dans leur capacité à modeler fondamentalement et essentiellement le socle d'une culture, mais au contraire à pister leur étoilement subtil en second plan, sous couvert de dérision ou de *ténuité*.[3] A partir du vêtement libéré, de l'étoffe qui ne moule plus aucun corps, ne voile plus rien, se forge une figure déréglée, à part, dont aucun projet discursif ou figuratif ne viendra plus figer le mouvement, un brouillage optique du cadre dont les apparitions/disparitions tendent à l'hallucination. L'autonomie et l'effacement de la figure marquent la fin de la circonscription, que l'étoffe dans sa chute vient subtilement souligner, une extériorisation qui éloigne corps et étoffe dans un mouvement réciproque d'expulsion. Dans la transformation du vêtement, du moule au drapé ou du *corps-froissé* au tissu inhabité, on observe la circulation d'une intensité, où l'espace pictural adopte ce qui marquait la limite déchue du corps, son vêtement, comme limite de/à la Représentation. En même temps qu'il destine ou projette le drapé vers la marge, le renvoyant à sa propre inutilité ou *vacuité*, il libère sa puissance formelle. Son exclusivité est tout à la fois son exclusion et la garantie de sa persistance.

Le *en trop* (*ghost track* pour nous), l'*en dehors* de l'œuvre (*par-ergon*) ouvre la possibilité formelle d'un *autrement*, qui garantit déjà les conditions de sa dispersion et de sa *survivance*. L'ornement ou du moins ce qui s'érige (ou choit, se dépose) comme tel, est ici, à plus d'un titre, une coïncidence, et son concept, une hallucination.

2. La popularité dont joui aujourd'hui l'œuvre de Gottfried Semper, nous invite à jouer la stéréotomie contre la tectonique, le Motif contre le Modèle, comme si, à sa périphérie, la pratique se trouvait revitalisée par des principes formels

plus essentiels (son essence?), plus fluides (son fluide anthropologique?)

On le sait, le principe de l'essence textile de l'Architecture énoncé par Semper, le *Prinzip der Bekleidung*, pose l'enveloppe, dans sa texturalité, comme primitive et antérieure à la structure. L'Architecture commence avec et par l'ornement, ou plutôt par la toile tressée, non pas par sa capacité à prodiguer un abri ou établir une clôture, mais par sa nature fibreuse même. Mark Wigley écrit à propos de la thèse de Semper:

The interior is not defined by a continuous enclosure of walls but by the folds, twists, and turns in an often discontinuous ornemental surface.[4]

La discontinuité de la surface tressée intervient ici à différents niveaux. D'abord elle suppose une définition d'une architecture à peu près indifférente à l'idée de circonscription spatiale. D'autre part, comme on le verra plus loin, elle évoque une forme heurtée, déchirée du projet. Il ne s'agit pas pour nous, en évoquant les textes de Semper, de faire référence à une forme ou formule originaire, un agencement premier entre structure et surface qui fixerait idéalement une juste répartition des rôles. Au contraire, faire basculer la dialectique structure/surface vers la primauté de l'ornement fige encore trop souvent celui-ci dans son rôle d'écran, de masque ou d'*intersecteur*.[5] Plus encore, il gèle le libre-jeu de son concept.

Le devenir-ornement de la surface tressée entame le deuil de la superposition du motif et de la structure, du *Werkform* et du *Kunstform*, car c'est bien le motif qui, structurellement, tient le drapé. Wigley le dit: "Strictly speaking, it is only the decoration that is structural."[6] Ce qui importe, dans la surface tressée, c'est la coïncidence d'un graphisme et d'une structure, le motif textile qui détermine sa tension interne propre et son comportement physique. Le plissé y surajoute un motif second, plus ample, distinct mais dépendant de sa structure première. Mais, du processus différentiel de la constitution de la surface même, nous tirons des enjeux structurels et formels qui, il nous semble, dépassent largement le débat *Werkform/Kunstform* sus cité.

3. Manuel De Landa[7] note qu'il est plus facile de modéliser numériquement un savoir verbal (comme "apprendre" à une machine à jouer aux échecs par exemple) qu'un savoir-faire, une technique de construction, d'élaboration manuelle qui suppose l'irrégularité, l'hésitation, l'essai, des logiques internes de compensations, des condensations hétérogènes et programmatiques, des assemblages décousus/recousus dans un souci de tenue générale de la surface. La technique de tressage évoque une telle complexité, de telles stratégies de finalisation. Elle renvoie au pouvoir et au produit d'une intelligence transformative, le processus mental et matériel commun à toute *tekhnê*.

Le corpus de gestes techniques associés à ce savoir, donné comme un ensemble de savoir-faire, résonne d'un basculement de la limite: art du tressage, du nœud, motifs torsadés, entortillés, oscillation du fil autour de l'étoffe, sont autant l'écho de figures inversées/renversées à l'infini, que de rêves d'espaces sans couture, sans jointure, dont le maillage absorbe, par sa densité, la linéarité de son contour.

Le tressage procède d'une économie générale, un principe différentiel de diminution des interstices, comme d'une illisibilité de sa logique constitutive. Au même titre que la tapisserie recouverte de motifs ornementaux au principe de tracé simple mais dont la répétition, la récurrence procède par complication progressive dans le temps et l'espace[8], la surface tressée tend à la perdition de la figure, à la fantomatisation du processus d'élaboration, à une spectro-graphie générale: une pulsation interne à la surface, graphique *et* structurelle, qui brouille la lisibilité de son projet, de ce qui, initialement programmé, en aurait permis le tressage. La condensation des nœuds, la saturation du tissage, aménagent une zone d'indiscernabilité ou d'indécidabilité dont il faudrait mesurer (cultiver?) l'amplitude.

A l'évidence de la régularité d'un tressage s'oppose la technique heurtée, la transformation de son projet qui est, à plus d'un titre, la trahison d'une mémoire, d'un énoncé préalable engagé dans les mouvements mnésiques de résurgences, récurrences, déchirures ou anamnèses. Le *savoir-tresser* en tant que *savoir-faire*, intègre organiquement l'incident à son déploiement temporel et spatial. C'est ce qui en définit son caractère éminemment événementiel. Il progresse par actualisations contextuelles successives. La surface tressée se fait en se dé-faisant, en contrariant, non pas systématiquement, mais de manière opportune, contextuelle, la forme programmée de ses nœuds.

On peut se demander dans quelle mesure ça n'est pas tant un certain aveuglement qui est programmé, une réduction du champ projectuel comme limite de pré-voyance, en même temps et à rebours d'une topologie nodale. Une stratégie qui imposerait un aller-retour entre l'œil et la main ou l'un fantomatiserait l'autre de manière contradictoire et compensatoire.

Voilà qui pourrait nous inviter à repenser l'(in-)actualité du "fait-main" et le devenir-fantôme de l'ornement à l'évidence de sa disparition, à l'époque de sa reproductibilité technique, au seuil de la Modernité. Ceci non pas dans le projet de sa réhabilitation, déjà bien entamée, mais dans la limite de son concept même. La réhabilitation de l'ornement *en tant qu'*ornement jette un voile sur la plus grande *activité* de ses formes, sa *puissance résiduelle* comme dit Didi-Huberman. Elle ignore la topologie nouée du *parergon* et l'énergie qu'il dissipe vers l'intérieur. L'éloge de la parure *comme* parure ne voit pas que s'y joue, à sa surface, un agencement de force qui vrille déjà programme/contexte/forme en un dispositif inouï.

1 Georges Didi-Huberman, *Ninfa Moderna, Essai sur le drapé tombé*, Gallimard, 2002

2 Jacques Derrida, *La vérité en peinture*, Flammarion, 1978, p.72

3 "(...) ce qui survit dans une culture est *le plus refoulé,* le plus obscur, le plus lointain et le plus tenace, de cette culture. *Le plus mort* en un sens parce que le plus enterré et le plus fantomal; *le plus vivant* tout aussi bien parce que le plus mouvant, le plus proche, le plus pulsionnel." Georges Didi-Huberman, *l'Image survivante, Histoire de l'Art et temps des fantômes selon Aby Warburg,* Editions de Minuit, 2002, p.154

4 Mark Wigley, *White Walls, Designer Dresses, The Fashioning of Modern Architecture,* MIT Press, 1995, p.11

6 Voile qu'Alberti conseillait au peintre de placer entre lui et la scène à représenter pour mieux en percevoir les contours.

6 Mark Wigley, Ibid, p.11

7 Manuel De Landa, *Les Philosophies de conception: le cas du logiciel de modélisation*, in *Verb, architecture boogazine,* ACTAR, 2002, p.134

8 E.H. Gombrich, *The sense of order – a study in the psychology of decorative art,* Phaidon Press Ltd, 1979, p.67

1.1 GOSTRACK

Biennale de Venise

Il est plus facile de créer un modèle numérique de connaissances verbales (comme "apprendre" à une machine à jouer aux échecs, par exemple) que de modéliser un savoir-faire ou un processus de fabrication, une technique d'élaboration manuelle qui implique l'irrégularité, l'hésitation, les essais et les erreurs, des logiques internes de compensation, dans une tentative de maintenir l'ensemble de la surface.

La technique du tressage évoque ce type de complexité où les mains interagissent avec la forme dans chacun des nœuds. Elle met en évidence la puissance et le produit d'une intelligence transformatrice, qui est un processus mental et matériel commun à tout type de technique.

L'art de tresser, de nouer, de torsader et de boucler les motifs, l'oscillation du fil autour du tissu génèrent une figure inversée et renversée incessante, comme des rêves d'espaces sans coutures, où le tissage perd dans la densité de la surface la linéarité de son contour.

Comme un tapis couvert de motifs ornementaux, la surface tissée utilise un mouvement simple, dont la répétition et la récurrence procèdent par complication progressive dans le temps et dans l'espace.

1.2 UNE CANOPÉE VIVANTE

Le nouvel Auditorium de Saint-Cyprien

Les arbres ont toujours été une source d'inspiration profonde pour moi. Leurs structures, bien qu'incroyablement complexes, émergent de simples mais puissantes règles génératives, évoluant de manière fluide dans le temps et l'espace. L'efficacité de leur architecture n'est pas dictée par une rigide répétition ou optimisation, comme dans l'ingénierie moderne, mais plutôt par la redondance et la différenciation, principes qui leur permettent de s'adapter, de croître et de prospérer.

Cette approche organique a guidé la conception du nouvel Auditorium de Saint-Cyprien, situé au cœur d'un parc ouvert. Ici, l'intérieur et l'extérieur se confondent, créant un environnement à la fois détendu et ouvert–une architecture qui invite au mouvement, à l'interaction et à la sérénité.

Inspirée par les silhouettes des sycomores, des acacias, des chênes et des peupliers qui peuplent le site, la forme de l'auditorium fait écho aux motifs de ramification complexes des cimes des arbres. Serero Architects a développé un script informatique personnalisé pour générer une façade composée d'éléments non répétitifs et non standard. Malgré son apparence apparemment irrégulière, la structure du toit est basée sur des règles géométriques simples, permettant une variation et une fluidité organique entre ses éléments.

Une architecture en continuité avec la nature

L'Auditorium Saint-Cyprien n'est pas conçu comme un objet architectural isolé, détaché de son environnement. Au contraire, il établit un dialogue continu avec le parc, intégrant le paysage dans le bâtiment lui-même.

Une double coque en béton crée l'impression d'une broderie de feuillage, percée de trous en forme d'œuf, rappelant la lumière tamisée filtrant à travers les feuilles des arbres. Cette enveloppe extérieure en forme de canopée fait office de couche protectrice, protégeant le hall et l'auditorium d'un ensoleillement excessif. L'enveloppe intérieure, composée de verre et de béton, régule le confort thermique, assurant une ventilation naturelle et un chauffage contrôlé. Les visiteurs entrent dans le hall en suivant un «galet» en béton niché sous ce feuillage architectural.

Une enveloppe réactive et performante

Le toit en forme de hangar agit comme une peau vivante, régulant activement l'atmosphère à l'intérieur de l'auditorium:

–Adaptation à la lumière naturelle –Tout au long de la journée, l'éclairage naturel inonde l'intérieur. À l'approche du soir, des lampes intégrées prennent progressivement le relais, assurant une transition en douceur entre la lumière du jour et l'éclairage artificiel.

–Ventilation passive et confort thermique – Le système de ventilation exploite le refroidissement passif en été, tout en retenant la chaleur en hiver, ce qui minimise la consommation d'énergie.

–Intégration des énergies renouvelables – Des panneaux photovoltaïques positionnés sur le toit orienté au sud génèrent une partie de l'électricité du bâtiment, assurant une contribution énergétique durable.

Au-delà de ses performances fonctionnelles, cette verrière brodée devient une signature architecturale distinctive, définissant l'identité de l'auditorium. Le jeu des oculi en béton et des panneaux alvéolaires translucides crée un effet visuel dynamique et changeant, renforçant l'idée de l'architecture comme un organisme vivant et respirant.

Un auditorium conçu pour l'excellence acoustique

L'intérieur de l'auditorium a été méticuleusement conçu pour obtenir une acoustique optimale, garantissant une expérience auditive immersive pour divers spectacles:

–Murs en lattes de bois avec traitements acoustiques – Les murs intérieurs, revêtus de lattes de bois et de matériaux acoustiques spécialisés, sont soigneusement adaptés pour accueillir à la fois des spectacles vivants et des projections cinématographiques, offrant une qualité sonore exceptionnelle.

–Plafond à la géométrie complexe pour la diffusion du son – La conception complexe du plafond améliore la diffusion du son, en assurant une distribution uniforme de l'audio dans tout l'espace, créant ainsi une expérience riche et immersive pour chaque spectateur.

Ce nouvel Auditorium de Saint-Cyprien est une architecture en symbiose avec la nature, qui emprunte à l'intelligence des structures arborescentes pour créer un espace fluide, adaptatif et riche sur le plan sensoriel. Plus qu'une salle de spectacle, c'est une extension organique du parc, où lumière, air, son et structure s'harmonisent pour redéfinir la relation entre l'architecture et son environnement.

L'EXPÉRIENCE DU SON
Villa Médicis

J'étais dans le noir. Une nuit noire, profonde et enveloppante, à la fois inquiétante et inconnue. Une obscurité différente de tout ce que j'avais connu auparavant. Bien que la nuit s'éloignait lentement, son épaisse obscurité restait intacte, inflexible. Je n'avais aucune lumière pour me guider. Aucun réverbère à l'horizon, aucune bougie, aucune fenêtre éclairée. En quittant mon atelier à la Villa Médicis, après une longue nuit de travail, j'avançais prudemment, laissant mes yeux s'adapter, essayant de naviguer dans le vide.

Pas à pas, je prenais conscience de mes propres pas de marche, de leurs échos qui résonnaient dans une cour invisible. J'écoutais attentivement, discernant la direction de chaque son, sa distance, sa décroissance. Dans mon esprit, je commençais à reconstruire l'espace autour de moi: la densité d'une haie, la proximité d'un mur, l'immensité d'une cour. Puis, le murmure subtil d'une fontaine d'eau, puis une autre, et une autre encore: un alignement de sources d'eau formant un axe invisible. Instinctivement, je le reliais au grand axe transversal de la Villa Médicis. Soudain, un silence profond s'étendait devant moi, un vide abyssal. Et là, émergeant de l'obscurité, je découvris la ligne d'horizon de Rome, ponctuée de dômes, se multipliant à l'horizon.

Nuit après nuit, quittant mon atelier à l'extrémité ouest du jardin, j'errais dans les profondeurs de cette noirceur. Pour la première fois, je faisais l'expérience de l'espace non pas par la vision, mais par le son.

J'appris à lire la résonance des murs monumentaux, l'écho d'une façade en travertin, la vibration d'une loggia en briques, la douce absorption des pins parasols. Chaque nuit, je cartographiais la Villa Médicis à travers sa présence acoustique. Je prenais plaisir à décoder son paysage sonore, me perdant pour retrouver mon chemin uniquement grâce au son.
C'est à ce moment que ma fascination pour le son a commencé—une révélation que l'architecture pouvait être comprise non seulement comme une discipline visuelle, mais aussi à travers sa capacité de résonance, d'empreinte sonore.

De la perception à l'expérimentation
À cette époque, je développais des dispositifs acoustiques capables de remodeler l'environnement sonore de la grande salle de réception de la Villa Médicis. Plutôt que de concevoir de nouvelles formes architecturales, j'ai choisi d'explorer le son lui-même—son mouvement, sa spatialisation, son interaction avec l'espace construit.

À l'aide d'un logiciel de simulation acoustique, j'ai étudié et analysé la propagation des ondes sonores, non pas simplement pour les simuler, mais pour décoder leur logique sous-jacente: comment les fréquences et les résonances pouvaient programmer l'organisation des matériaux, définir de nouveaux principes structurels et générer des configurations spatiales intelligentes répondant aux exigences acoustiques.
Traditionnellement, les dômes ont été des solutions architecturales pour couvrir de vastes espaces, leurs géométries étant façonnées par les contraintes de résistance des matériaux. Mon séjour à Rome m'a offert la liberté de remettre en question cette tradition, d'imaginer un dôme qui transcenderait son rôle structurel pour devenir un dispositif acoustique dynamique.

Le dôme acoustique à configuration variable
En collaboration avec un mathématicien italien, un acousticien et un programmeur, nous avons développé une nouvelle façon de visualiser le comportement acoustique, une méthode qui nous a permis de générer des champs acoustiques comme fondements projectifs de l'architecture elle-même.

Installé dans le hall de réception de la Villa Médicis, haut de 14 mètres, le dôme acoustique à configuration variable reflète le plafond voûté de la pièce. Cette structure expérimentale, suspendue au-dessus de l'espace, est conçue pour ajuster et transformer le paysage sonore de la salle en temps réel.

En modifiant sa position et son degré de pliage, le dôme remodèle activement la propagation et la diffusion du son. Ici, l'espace n'est pas passif, il devient un participant à la performance, un médium interactif. À chaque concert, à chaque conférence, à chaque voix qui remplit la salle, la géométrie de la pièce s'adapte, modifiant les conditions acoustiques en réponse à l'événement et à son public.

Le dôme est composé d'une surface pavée, où un système itératif de plis génère une harmonie évolutive entre la musique et l'architecture. Sa forme est conçue comme une convergence entre le calcul acoustique numérique et le déploiement physique des matériaux.

Plutôt que de définir une forme fixe, le projet embrasse la variabilité, intégrant les principes mêmes de la transformation dans sa géométrie. Il opère à l'intersection de l'ornement et de l'acoustique, permettant un échange actif entre l'architecture, la musique et la perception.

La coupole acoustique crée une relation fluide et dynamique entre l'espace, l'œuvre, le public et son environnement immédiat. Il ne s'agit pas simplement d'une installation architecturale, mais d'un instrument de résonance, d'une structure qui écoute autant qu'elle façonne le son qu'elle reçoit.

1.3 DÔMES ACOUSTIQUES À GÉOMÉTRIE VARIABLE

Villa Médicis – Rome

La musique, phénomène à la fois tangible et intangible, sert de fondement à cette intervention spatiale au sein du grand hall de réception de la Villa Médicis.

Suspendu dans cet espace majestueux de 14 mètres de haut, le dôme acoustique à configuration variable est un dispositif expérimental qui reflète le plafond voûté de la pièce. Conçue pour ajuster dynamiquement sa géométrie, l'installation transforme les propriétés acoustiques de l'espace, modifiant à la fois son environnement sonore et sa réponse de réverbération.

En modifiant sa position dans le volume de la salle et en ajustant son degré de pliage, le dôme remodèle la façon dont le son se propage et se diffuse. Ici, l'espace n'est plus un conteneur passif pour la musique, il devient un média interactif, où les compositions musicales sont intégrées dans une forme adaptable. À chaque représentation, à chaque public et en temps réel, la géométrie de la salle elle-même évolue pour entrer en dialogue avec la musique.

La nature de l'événement, qu'il s'agisse d'un concert de musique de chambre, d'une conférence ou d'une représentation vocale, influence directement le calcul de l'espace qui régit les mouvements du dôme. À mesure que la coupole acoustique se plie, se déplie et s'adapte, la relation entre l'architecture et le son est continuellement redéfinie. La tessellation de sa surface, combinée à un mécanisme de pliage itératif, permet une synchronisation dynamique entre la musique et la forme spatiale.

Ce projet ne cherche pas à définir une forme fixe, mais plutôt un système adaptatif, intégrant les principes mêmes de sa transformation dans sa géométrie. Il existe à l'intersection de la forme ornementale et de la performance acoustique, créant un dialogue entre la structure et la propagation du son. Le dôme établit une relation flexible et évolutive avec l'architecture existante, l'œuvre musicale et le public, brouillant la frontière entre l'espace, le son et la perception.

1.4 CENTRE CULTUREL ROBERT DOISNEAU

Meudon-la-Foret

Ce projet s'inspire de la nature, et plus particulièrement de l'arbre, non pas dans sa simple forme symbolique, mais précisément dans son rapport profond à l'architecture, dans les relations qu'il entretient entre structure et enveloppe, entre redondance et optimisation, entre force et fragilité, entre singularité et diversité. L'interaction entre l'espace naturel de la forêt et les volumes minimalistes des bâtiments existants conçus par l'architecte Fernand Pouillon ont constitué l'axe principal de notre projet. Le bâtiment s'implante dans la continuité de la trame orthogonale du quartier tout en développant une relation ouverte et multiple avec l'espace au sol. Le projet se déploie autour du noyau central que constitue la salle de spectacle de 600 places. Son enveloppe périphérique se soulève au-dessus du hall pour révéler les espaces intérieurs et les zones de circulation. Ce dispositif, en créant des fenêtres sur toute la périphérie du bâtiment, permet d'éviter l'effet «boîte noire» de la salle de spectacle plantée au milieu de l'environnement urbain. A l'opposé de l'unique module de fenêtre utilisé par Pouillon pour tous les logements du quartier, nous avons conçu un système de baies vitrées de formes libres, dont la géométrie est dérivée de la géométrie des branchages des arbres.

Ce projet est conçu de l'intérieur vers l'extérieur, proposant un travail sur le sol et sur le toit qui cadre les espaces sans les fermer. Le bâtiment est en porte-à-faux au niveau du hall d'accueil et du parvis de manière à créer de larges surfaces vitrées protégées du soleil et de la pluie. En périphérie, courent sur deux niveaux les espaces d'accueil et de rencontre, le hall, les bureaux et les ateliers artistiques.

Tous les espaces de circulation et d'accueil sont largement baignés de lumière naturelle, soit par le mur rideau du hall d'accueil, soit par les sheds qui scandent la toiture des ateliers et jouent un rôle important dans la définition des ambiances lumineuses à l'intérieur du bâtiment. De par la variation de taille des éléments qui la constituent, l'ambiance lumineuse est proche de celle d'un sous-bois. L'éclairement à l'intérieur du hall principal change ainsi à chaque heure de la journée. En fin de journée, des lampes installées dans ces sheds et reliées à des cellules photosensibles compensent progressivement le manque de lumière naturelle. Les bureaux et les salles de travail sont éclairés par des baies multiples de formes libres qui permettent, dans chaque lieu, d'avoir des vues vers l'extérieur et les espaces paysagers.

Le bâtiment est recouvert d'une enveloppe en béton architectonique de couleur claire qui intégrera des gravillons de calcaire de Meudon dans sa composition. Conçue comme une peau intelligente de grande performance, la toiture est un élément important de ce projet car elle intègre des dispositifs de traitement des ambiances lumineuses et de contrôle thermique.

1.5 EIFFEL DNA

Restructuration du 3ème plateforme de la tour Eiffel - Paris 2012

«... les lois authentiques de l'énergie ne sont pas toujours conformes aux lois secrètes de l'harmonie...»

Gustave Eiffel

La tour Eiffel, symbole éternel de Paris, souffre du poids de son succès. Depuis son inauguration, le nombre de visiteurs atteignant son sommet n'a cessé d'augmenter, poussant la structure à ses limites de capacité. Aujourd'hui, 6,5 millions de visiteurs par an subissent des temps d'attente allant de 35 minutes à plus d'une heure pour accéder à son sommet. La géométrie effilée de la tour aggrave encore la situation, car la surface au sol diminue avec la hauteur, ce qui crée des problèmes d'encombrement et de gestion des foules.

Pour célébrer le 120e anniversaire de la tour, nous avons imaginé un projet spéculatif qui repense ses espaces d'accueil et d'accès publics. La proposition introduit une prolongation horizontale temporaire du troisième étage, conçue pour améliorer à la fois la circulation des visiteurs et l'expérience du panorama époustouflant à 360 degrés de Paris.

Lorsque Gustave Eiffel a conçu la tour, il l'a vue comme une plate-forme d'exploration scientifique, capable de supporter divers instruments pour étudier la gravité et la pression du vent. La structure a été dimensionnée pour supporter un poids supplémentaire, une prévoyance qui s'est avérée précieuse pendant la Première Guerre mondiale lorsque la tour a abrité des antennes pour les transmissions radio nationales. Les analyses structurelles confirment que la tour Eiffel est une structure hautement hyperstatique, conçue pour des charges supérieures à celles qu'elle supporte actuellement.

Tirant parti de cette résilience structurelle, notre projet propose une extension légère en carbone-kevlar, temporairement boulonnée à la dalle du troisième étage sans altérer la charpente d'origine. Cette intervention permettrait d'étendre la surface utile de 280 m² à 580 m², améliorant ainsi considérablement l'accessibilité. L'extension se compose de deux plateformes supplémentaires, parfaitement intégrées à la terrasse du troisième étage et au niveau supérieur, autrefois l'appartement privé de Gustave Eiffel, préservant ainsi l'esprit de l'importance historique de la tour.

La conception est paramétrique, inspirée des principes structurels de Gustave Eiffel, à savoir les poutres de contreventement tridimensionnelles. À l'aide de scripts génératifs, nous avons développé un système de ramification qui émerge de la structure existante de la tour, faisant écho à son ADN en treillis distinctif. Ancré dans le cube de 10 x 10 mètres au sommet, le script génère trois couches structurelles entrelacées, formant un tissage complexe mais très efficace. Ce motif non répétitif améliore les performances structurelles, s'écartant de l'ingénierie moderne conventionnelle, qui repose sur la répétition et l'optimisation. Cette approche explore plutôt un modèle alternatif de conception haute performance, où la variation organique renforce le système global.

En prolongeant temporairement les niveaux supérieurs de la tour Eiffel, ce projet redéfinit non seulement sa capacité, mais réinterprète également son héritage, en faisant le lien entre l'ingéniosité historique et l'exploration architecturale contemporaine.

1.6 MODULATIONS

Jardin De Métis

L'architecture a longtemps été considérée comme la création d'un ordre géométrique et donc artificiel, par opposition au jardin comme ordre naturel. Aujourd'hui, cette distinction s'estompe de plus en plus, à une époque où les pratiques architecturales s'inspirent des processus naturels, des modèles évolutifs et des modes de transformation. Il est également évident que l'architecture a longtemps défini un cadre pour le jardin. Il s'agit désormais d'une pratique plus large qui, au-delà des débats sur la forme et la fonction, définit avant tout un système d'ambiances, un "climat" continu qui englobe les échelles territoriale, paysagère, urbaine, architecturale et corporelle.

La pratique du paysage s'est longtemps fondée sur le principe de mise en scène de la nature par la régulation et l'orthonormalisation des axes et des lignes. Les éléments naturels étaient calibrés et "célébrés" dans une relation connue de l'homme à son environnement. Aujourd'hui, la ville et le jardin sont de moins en moins deux entités distinctes. La pratique du paysage favorise l'émergence de nouveaux environnements par le développement de systèmes écologiques graduels et la transformation progressive des éléments contextuels et spécifiques.

Le jardin, métaphore d'un paradis originel et lieu de recréation de la nature, est devenu un champ de création spécifique qui se nourrit de méthodes conceptuelles développées, entre autres, par des influences artistiques et architecturales. Il entretient désormais une relation plus ambiguë entre l'horizon architectural et le milieu naturel. Ce travail quasi instantané sur l'espace se définit ainsi de plusieurs manières.

Le jardin éphémère est un mode d'expression contemporain de l'espace basé sur la consommation de masse, périssable dans le temps (un changement annuel), mais dont les concepts peuvent être adaptés ailleurs. Il n'est pas l'expression spécifique d'un médium, mais participe d'un réseau global d'échanges entre différents créateurs.

Le jardin véhicule une certaine idée de la nature où les "éléments bruts" sont contraints de dialoguer dans un espace restreint. Le jardin se définit par les relations qu'il établit, dans un rapport de friction entre éléments naturels et artificiels. C'est un espace laboratoire, un lieu d'expérimentation.

L'intention n'est pas de rechercher une quelconque harmonie ou homogénéité, mais un dialogue, une nouvelle articulation, où chaque visiteur expérimente sa propre manière de traverser l'espace. Ainsi, notre jardin ne prend sens et ne peut être compris que lorsqu'il est mis en «action» par le visiteur.

C'est ce que nous avons exploré à Grand Métis en 2008, lors de la réalisation du jardin temporaire "Modulations" avec le paysagiste Philippe Coignet. Inspirés par la simplicité d'une feuille de fougère, nous avons déployé son mode de croissance et sa formation géométrique à plus grande échelle pour créer des écrans en acier inoxydable. Ces grandes bandes de métal ont été développées en simulant la croissance des plantes, avec des paramètres de croissance et de génération de feuillage analysés pour leur capacité à refléter et à dissimuler l'espace environnant. La géométrie des bandes d'acier reproduit celle de la feuille elle-même, reflète et diffracte la lumière de cet espace, de manière presque mimétique à la plante.

La complexité des phénomènes naturels impliqués dans le paysage est maintenant mieux comprise. La recherche biologique et informatique permet de comprendre les règles et les processus de la formation naturelle, de les détourner et de s'en inspirer pour créer de nouvelles formes.

Le motif végétal est ici le prétexte à l'invention d'un nouveau jardin, à la fois réalité simulée et surface évanescente qui oblige le visiteur à se déplacer et à découvrir. Ce projet engage un renversement des limites du jardin à l'intérieur de lui-même par un dispositif visuel. Ce n'est que lors du déplacement de notre corps dans le jardin que les reflets et les jeux de lumière sur cette surface rendent lisible cet espace en transformation: ces changements expriment alors la notion de temporalité.

1.7 LEARNING AND RESEARCH CENTER
Issy-les-Moulineaux

La restructuration du centre de formation CNFPT à Issy-les-Moulineaux a pour objectif de créer un lieu d'innovation pédagogique ouvert. Composé de 22 salles de formation, d'amphithéâtres, de salles numériques, d'espaces de co-conception et de bureaux, le projet consiste à décloisonner l'espace de la salle de cour en ouvrant de multiples voiles porteurs en béton au sein du bâtiment. Cette stratégie permet de laisser pénétrer la lumière naturelle au cœur du bâtiment et d'offrir de nouveaux espaces de formation alternatifs.

Nous avons donc considéré le bâtiment comme une éponge que nous avons évidé en créant des nouveaux espaces libres et "informels" associés aux espaces de formation traditionnels. De cette manière l'enveloppe extérieure et les aménagements intérieurs s'accordent pour éclairer les espaces du bâtiment en son centre. Ces "poches intérieures" de lumière sont associées à un travail sur les cloisons intérieures acoustiques qui sont transparentes, et semi-transparentes par endroits.

Les couloirs entre les salles ont été supprimés au profit d'un immense espace tiers-lieu sur trois niveaux. À la fois espace d'accueil, d'échange, mais aussi d'autoformation, ces lieux libèrent les usages et créent un dispositif pédagogique innovant.

Au rez-de-chaussée, cet espace central est un learning center associé à un fonds documentaire et une médiathèque. Au première étage, cet espace est un espace de convivialité, à la fois cafétéria et espace de coworking, au second étage, c'est un atelier de pratiques collectives qui s'ouvre sur un potager et une terrasse jardin. Dans ce projet, les couloirs et les escaliers, espaces généralement dédaignés, deviennent ainsi des lieux actifs dans la production de savoirs et dans l'échange de connaissances.

Notre travail n'a pas cherché à imposer une esthétique au bâtiment existant. C'est plutôt l'inverse qui s'est produit. Nous avons travaillé sur l'existant par touches successives afin de déshabiller ces espaces et de rendre visible leur structure interne: les dalles de béton préfabriquées, les poteaux, les gaines de ventilation, le chemin de câbles ont été dépouillés de leur revêtement pour les considérer comme des éléments intégraux de l'expérience utilisateur.

2. ENERGIE DÉPLOYER DES PRATIQUES ARCHITECTURALES BAS CARBONE

Notre philosophie définit la question énergétique comme indissociable de la question architecturale. Energie et architecture sont intrinsèquement liées, formant un objectif commun qui façonne à la fois le design architectural du projet et son approche environnementale.

Que ce soit dans le processus de conception du projet, dans le choix du mode constructif, dans le cycle de vie des matériaux ou encore le choix du système énergétique du bâtiment, le travail de l'architecte est aussi celui d'un énergéticien. Il conçoit et met au point des dispositifs de consommation, mais aussi de production et de renouvellement des énergies, qu'elles soient techniques et physiques, humaines ou matérielles. Dans chacun de nos projets, nous avons cherché une dialectique au sein du projet architectural entre dispositifs bioclimatiques, issus de l'étude d'architectures originelles, rurales ou vernaculaires, avec des technologies plus avancées d'adaptation ou d'optimisation permettant une interaction entre l'homme et les espaces de nos projets. (Enveloppes architecturales adaptatives, routines et algorithmes liés aux usages, adaptation de la ventilation naturelle et contrôle solaire).

Avec la mise en place de nouveaux objectifs européens de réduction énergétique et d'optimisation des villes, les pratiques d'architecture bas carbone prennent enfin de l'ampleur. C'est une réflexion globale sur le bilan environnemental global et le cycle de vie d'un bâtiment que les équipes de maîtrise d'œuvre vont développer: une architecture bioclimatique, optimisant l'utilisation des ressources naturelles et la mise en place de nouveaux circuits bio et géo-sourcés, en pensant le bâtiment en relation avec un réseau énergétique partagé ou autonome.

Les trois premiers projets présentés ici, la médiathèque de Bayeux, la bibliothèque et l'auditorium du Campus universitaire d'Amiens et le centre de recherche Climat Planète, sont trois réalisations significatives du développement de cette réflexion.

Le projet d'Amiens a permis de mettre au point un système d'enveloppe architecturale de contrôle solaire par des écailles à angles variables–avec un réglage de leur orientation par saison—issu d'un modèle bio inspiré telle que celui du cône de pin. La médiathèque de Bayeux, est issue d'un travail ou les vitres et murs-rideaux, les écrans intérieurs filtrants et les débords de toiture sont précisément calculés pour magnifier la transparence du bâtiment avec une protection solaire et thermique optimales. Le projet Climat Planète, un centre de recherche sur le climat à Grenoble, a visé à rechercher un équilibre énergétique parfait, entre température de nappe phréatique, rafraîchissement d'un Data center et régulation thermique de bureaux et laboratoires des chercheurs.

Dans le cas du parc métropolitain Hellenikon à Athènes, un projet sur la restructuration du site gigantesque de l'ancien aéroport d'Athènes, les réflexions de l'équipe se sont inspirées des processus naturels de recolonisation d'espaces, pour les déployer sur une infrastructure aéroportuaire en limitant les énergies utilisées et le coût du projet. Nous nous sommes en particulier intéressés à la gestion de l'eau sur le site et à la manière dont un aménagement paysager de très grande échelle pourrait permettre de canaliser la force des inondations fréquentes dans cette zone, dûes à l'artificialisation des sols, par un profilage topographique adapté à la collecte et le stockage des eaux de ruissellement, et initier un nouveau cycle de l'eau vertueux.

Le dernier projet pour l'université d'Orléans présente une réflexion singulière sur la rénovation énergétique des bâtiments, un chantier mondial de transformation et d'adaptation des bâtiments aux changements climatiques qui concernerait plus de 60% du parc mondial de bâtiments. Ici, nous avons associé un projet de transformation pédagogique des amphithéâtres par des systèmes de travail en grappe, avec une isolation thermique extérieure avec bardages en résine de béton, associé avec une démarche originale d'accueil de la biodiversité. Avec le partenariat du muséum d'histoire naturel de Bourges, nous avons identifié la présence d'une espèce locale de chauve-souris sur le bâtiment, la pipistrelle, que nous avons pu relocaliser après travaux par un système de nichoirs intégrés à la façade et à son isolation.

Un modèle de simulation pour repenser la question de l'énergie

Pourrions-nous proposer une nouvelle approche du projet architectural en associant les outils d'analyse et de calcul de phénomènes physiques à l'expertise d'un architecte en milieu urbain?

L'essai présenté ici est issu d'une recherche, à mi-chemin entre l'architecture et l'ingénierie, et s'appuie sur un modèle de simulation numérique, qui permet une analyse du comportement du bâtiment en phase de conception. Ces informations constituent un environnement de conception de projet visant à "profiler" le bâtiment, selon une méthode analytique et performative plutôt par une simple approche subjective de design architectural.

Le travail et la recherche s'appuient sur un double corpus de connaissances: la conception architecturale et la mécanique des fluides, permettant de sortir l'architecture de son isolement entre les arts et l'ingénierie et de proposer une fertilisation croisée de ces disciplines. L'approche est donc transdisciplinaire et vise à étudier l'impact du vent dans les zones hyper-localisées de grands bâtiments, et à informer la conception architecturale du changement de flux d'air inévitable à la périphérie de son enveloppe.

Cette analyse révèle que tout bâtiment, tout mur, génère par sa présence sur un terrain, un différentiel de pression d'air sur ses parois externes, lié au déplacement des couches d'air environnantes. Cette étude a montré que la volumétrie et la position d'un bâtiment dans son contexte peuvent ainsi être "profilées" et "orientées" pour une augmentation et un contrôle de la vitesse du vent en sa périphérie.

Cette méthode de conception de bâtiments pour les architectes et les ingénieurs permettrait d'intégrer des dispositifs de captation d'énergie tels que micro-éoliennes en les intégrant aux enveloppes de leur bâtiments. Des structures de type "aérofoils", de type aile portante peuvent stabiliser localement les turbulences du vent et augmenter ponctuellement la vitesse du vent.

L'amélioration de la conception bioclimatique des bâtiments portée par cette recherche, combinée aux politiques environnementales mondiales de réduction des gaz à effet de serre, permet l'émergence d'un écosystème de partage de l'énergie.

L'émergence de l'Internet des Objets (IOT) permet aux différents composants du bâtiment d'interagir entre eux, mais aussi avec les utilisateurs et les systèmes de gestion des bâtiments. Il s'agit d'une infrastructure émergente, l'Internet de l'Énergie, qui vise à proposer une organisation pour le partage et l'échange d'énergie au niveau local. Ce nouveau paradigme projette une vision d'environnements urbains intelligents où les ressources, la production et la consommation d'énergie seraient hyper-localisées, voire autonomes, avec un impact environnemental vertueux.

2.1 SCALES OF LIGHT

Bibliothèque de l'IUT d'Amiens

Situés au cœur du campus de l'université d'Amiens, un centre dynamique de la vie étudiante, la nouvelle bibliothèque et l'auditorium constituent un pôle essentiel pour l'enseignement, la recherche et les échanges interdisciplinaires. Conçu comme un prolongement naturel du paysage, le projet s'intègre harmonieusement à la topographie du site, créant un dialogue entre l'espace bâti et l'environnement.

Le projet s'articule autour de deux plateformes, toutes deux alignées sur le jardin central de l'université: la bibliothèque au sud et l'auditorium au nord, dont les gradins suivent la pente naturelle du terrain. Entre ces deux espaces, un hall d'entrée sert de principal point de convergence, donnant accès aux deux volumes. Un escalier, discrètement placé dans l'espace entre les deux bâtiments, relie directement le hall d'entrée au niveau inférieur, renforçant la fluidité des déplacements entre les différentes zones.

La lumière naturelle joue un rôle fondamental dans l'atmosphère de la bibliothèque. Situé au cœur de la structure, l'espace est baigné d'une lumière douce et uniforme grâce aux toits en appentis orientés au nord qui s'étendent sur toute la longueur du bâtiment. Cet éclairage zénithal crée un environnement ouvert et accueillant, transformant la bibliothèque en un lieu d'échange, de partage des connaissances et de rencontres interdisciplinaires.

L'auditorium, construit avec des murs en béton léger, bénéficie de hautes fenêtres continues sur toute sa longueur, permettant à la lumière naturelle de pénétrer dans l'espace. Ces fenêtres peuvent être complètement obscurcies à l'aide de stores intégrés, garantissant des conditions optimales pour les conférences, les projections et les présentations.

Conçue sur une plate-forme libre avec un minimum d'interruptions structurelles, la disposition ouverte de la bibliothèque est soutenue par de minces colonnes intermédiaires de 10,80 mètres. Un système de grille modulaire de 1,80 m et 7,20 m organise la structure, l'éclairage, la ventilation et la répartition du mobilier, créant un intérieur intuitif, lisible et très flexible. Cette clarté de l'organisation améliore l'expérience de l'utilisateur, facilitant l'accès intuitif aux livres et aux ressources tout en équilibrant les espaces d'étude communs et calmes. La transition entre les grands espaces de réception et les zones de travail plus petites et plus intimes favorise une atmosphère propice à la collaboration et à la concentration individuelle.

La façade du bâtiment fonctionne comme une peau intelligente, gérant activement la lumière, les vues et le contrôle climatique. Inspirés par les écailles d'une pomme de pin, qui s'ouvrent et se ferment en fonction de l'humidité, les brise-soleil en bois s'adaptent à l'orientation du bâtiment, créant une interaction dynamique entre transparence et protection. Ces éléments verticaux changent progressivement d'angle le long de la façade, s'adaptant aux besoins spatiaux et environnementaux de l'intérieur. Positionnés de manière à optimiser l'exposition au soleil de l'ouest, ils offrent un filtrage supérieur de la lumière solaire rasante, assurant à la fois le confort thermique et la clarté visuelle. Grâce à son intégration topographique, sa stratégie de lumière du jour et sa façade bio-inspirée, la bibliothèque et l'auditorium d'Amiens redéfinissent la relation entre l'architecture, le paysage et le savoir, offrant un espace harmonieux et durable pour l'apprentissage et l'échange.

2.2 TISSER LE VERRE ET ACIER

Médiathèque de Bayeux

Située en bordure du périphérique qui encercle le cœur historique de Bayeux, avec son labyrinthe de rues étroites et sinueuses, la nouvelle médiathèque de Bayeux s'intègre parfaitement dans le riche tissu architectural et historique de la ville. Situé sur un site dégagé, le bâtiment fait le lien entre le centre médiéval et les futures zones d'expansion de la ville, offrant une vue imprenable sur la cathédrale. Ce cadre exceptionnel exigeait une réponse architecturale sensible. Le design privilégie la transparence, créant un bâtiment-paysage qui sert de vitrine urbaine vivante. Ouvert et accueillant, le projet favorise un dialogue fluide entre l'intérieur et l'extérieur, avec un patio central et des terrasses de lecture extérieures côté sud, prolongeant les espaces et les activités de la bibliothèque à l'air libre.

La lumière naturelle est un élément clé de la conception de la bibliothèque. Chaque espace est éclairé par des sources lumineuses variées, soigneusement orchestrées pour créer de la clarté et une lisibilité spatiale. Des puits de lumière ponctuent le toit végétalisé, apportant la lumière du jour au plus profond des intérieurs, tandis que le patio central canalise la lumière au cœur du bâtiment, favorisant une atmosphère lumineuse et accueillante. Une maille métallique perforée sur la façade nord adoucit la lumière entrante, assurant une diffusion douce dans les zones de lecture. Il en résulte une bibliothèque hautement transparente et ouverte, où les utilisateurs se dirigent intuitivement vers les livres et les ressources. L'entrée nord-est, légèrement en retrait de l'intersection urbaine, accueille les visiteurs dans un parvis calme et accueillant, à l'écart de la circulation environnante.

L'histoire de Bayeux est indissociable de son artefact le plus célèbre, la Tapisserie de Bayeux, également connue sous le nom de Tapisserie de la Reine Mathilde. Tissée et brodée au XIe siècle, ce chef-d'œuvre a directement inspiré la façade nord de la médiathèque. Tout comme la tapisserie a été créée point par point, fil par fil, la façade nord de la bibliothèque incarne cet esprit artisanal. Côté boulevard de la Vacherie, toute la surface vitrée est protégée par une "peau filtrante" composée de tubes verticaux teintés. Ces éléments font référence aux teintes naturelles de la laine de la tapisserie—beige, marron, vert bronze, bleu-noir, bleu profond avec des reflets jaunes—brouillant la frontière entre le passé et le présent. Cette façade brodée filtre à la fois les vues entre l'intérieur et la rue et régule les niveaux de lumière, enrichissant ainsi l'ambiance des espaces de lecture.
La peau filtrante brodée fonctionne en tandem avec le toit en surplomb côté sud, assurant un confort visuel et une luminosité contrôlée à l'intérieur de la bibliothèque. Cette enveloppe adaptative joue un rôle essentiel dans la régulation thermique du bâtiment. La façade s'ouvre par des baies vitrées sur les côtés est, sud et

ouest, équipées de stores à enrouleur réglables, permettant aux utilisateurs de moduler l'apport de lumière et de chaleur. Le toit végétalisé, conçu comme une cinquième façade, est ponctué de puits de lumière qui améliorent la pénétration de la lumière du jour tout en permettant une ventilation naturelle pendant les mois de mi-saison. Cette synergie entre architecture, climat et histoire transforme la Médiathèque de Bayeux en un centre culturel lumineux et vivant, profondément ancré à la fois dans le patrimoine et l'innovation contemporaine.

2.3 LABORATOIRES DE RECHERCHE "CLIMAT PLANÈTE"

Le Centre de recherche Climat Planète est conçu comme un double bâtiment, prolongeant le laboratoire CERMO existant sur le campus de l'Université Grenoble Alpes (UGA) à Saint-Martin-d'Hères. Faisant écho à l'ADN architectural de son prédécesseur, la nouvelle structure s'articule autour de circulations verticales, avec une charpente rythmée et un plan de sol ramifié qui organise les espaces en volumes perpendiculaires. Plus qu'une simple extension, Climat Planète est conçu pour rassembler une communauté de chercheurs, favorisant les échanges grâce à son infrastructure de recherche, son centre de données et sa salle de virtualisation informatique.

À la jonction des deux bâtiments, un volume entièrement vitré révèle trois niveaux de plateformes suspendues, abritant des espaces d'accueil, des salles de réunion et des espaces de conférence. Ce pôle sert à la fois d'espace de travail collaboratif et de vitrine de l'activité de recherche, offrant une visibilité sur la dynamique interne du bâtiment. Afin de réguler la lumière et de maintenir le confort thermique, les façades est et ouest sont protégées par un treillis en acier inoxydable, qui filtre la lumière du soleil tout en préservant la transparence.

Les étages de bureaux sont entièrement ouverts et structurés autour d'une zone de circulation centrale élargie qui fait office de couloir de service et intègre des espaces de réunion informels, des zones d'attente et des systèmes de ventilation naturelle. Cette disposition favorise la flexibilité et l'adaptabilité, essentielles dans un environnement axé sur l'innovation.

Conçu pour anticiper l'évolution rapide de l'informatique quantique et numérique, le centre de données est conçu comme un grand espace ouvert avec une structure surélevée, permettant une reconfigurabilité maximale. Une salle de supervision et un espace de préparation des machines créent une interface sécurisée avec la plate-forme principale, tandis qu'une salle technique à double hauteur accueille les systèmes de ventilation et de refroidissement. L'entrée d'air frais est gérée par des persiennes en aluminium thermolaqué, assurant une régulation thermique efficace sur le plan énergétique.

L'extérieur du bâtiment est enveloppé d'une enveloppe de panneaux d'aluminium architecturés, recouvrant un système d'isolation thermique d'origine biologique. Un ensemble de cadres structurels accentue la géométrie du bâtiment, avec des angles et des transparences finement ajustés pour répondre au traitement solaire et à l'orientation de chaque façade. Le contrôle solaire est aussi assuré par des brise-soleil verticaux fixes en aluminium sur les façades est et ouest, complétés par des étagères lumineuses positionnées au-dessus des fenêtres pour refléter la lumière naturelle du jour profondément dans les espaces de bureau, ainsi que des stores motorisés pour l'adaptation dynamique des conditions d'éclairage intérieur.

Fondé sur les principes bioclimatiques, Climat Planète intègre un système intelligent de gestion de l'énergie, assurant une interaction parfaite entre le bâtiment et son environnement.

Une stratégie de ventilation naturelle couvre 90 % des besoins de renouvellement de l'air, minimisant ainsi la consommation d'énergie tout en tirant parti de l'inertie thermique des sols en béton pour réguler les températures intérieures. En été et à la mi-saison, le flux d'air est déterminé par les différences de température entre l'air ombragé au niveau du sol et la cheminée solaire sur le toit, ce qui améliore l'extraction naturelle de l'air. En hiver, un système de ventilation à double flux haute performance avec récupération d'énergie assure l'efficacité thermique.

La présence de réserves d'eau souterraine permet la mise en place de puits géothermiques, puisant l'eau à 11 °C pour réguler les conditions climatiques intérieures. Ce système permet non seulement de refroidir les serveurs et de maintenir le confort dans les bureaux, mais contribue également au réseau hydraulique du système de chauffage/refroidissement par le sol. L'intégration d'un système passif de chauffage/refroidissement par le sol, fonctionnant sur la circulation d'eau à basse énergie, réduit encore la demande énergétique.

Afin d'optimiser l'utilisation de la lumière du jour tout en contrôlant l'apport solaire, le bâtiment est équipé d'étagères à lumière qui diffusent la lumière naturelle dans les espaces de travail tout en assurant un ombrage passif sur les façades sud, est et ouest. Les salles de serveurs et les espaces informatiques à haute densité sont refroidis par un système de refroidissement naturel direct qui injecte de l'air extérieur filtré à l'avant des serveurs, expulsant efficacement la chaleur excessive vers l'extérieur.

Avec sa conception bioclimatique et ses stratégies passives avancées, Climat Planète est le premier bâtiment à énergie positive du campus de l'UGA. Il repousse non seulement les limites de l'architecture durable, mais redéfinit également l'environnement de recherche, en créant un espace où la science, l'innovation et la responsabilité environnementale convergent.

2.4 PARC MÉTROPOLITAIN HELLENIKON
Athènes

En refaisant surface aux traces des anciens cours d'eau qui traversaient autrefois le site —effacés lors de la construction de l'aéroport— notre projet rétablit les liaisons transversales entre les deux autoroutes qui bordent la zone. Ces "paysages doux" remodèlent subtilement la topographie pour créer des corridors naturels pour la végétation, les sentiers et les eaux de pluie, intégrant ainsi de manière harmonieuse les processus écologiques dans le tissu urbain.

La conception du parc s'articule autour de l'interaction de trois stratégies clés, générant un nouveau terrain qui mêle les éléments existants à une vision avant-gardiste de l'espace public.

Les "Hardscapes" tirent parti de l'échelle monumentale des anciennes pistes de l'aéroport, les traitant comme des "objets trouvés", des structures réutilisées pour définir les principaux axes visuels et de circulation du parc. S'inspirant de la "grande perspective" d'André Le Nôtre à Versailles, ces vastes surfaces de béton sont préservées en tant que boulevards piétonniers intérieurs, offrant un contraste saisissant entre patrimoine industriel et renouveau naturel. La piste Est/Ouest s'étend vers la mer, enjambant l'autoroute pour établir une liaison directe avec le littoral et ses plages. Les pelouses entre les pistes sont réinventées en une série de plateformes de détente, de jeu et de contemplation, transformant la géométrie rigide de l'aéroport en un paysage public accueillant et dynamique.

Les "sofscapes" suivent les schémas d'irrigation naturels du site, créant un réseau qui relie la ville et le parc grâce à une stratégie intégrée de gestion de l'eau. Pendant les périodes de fortes pluies, l'eau est canalisée, retardée et temporairement stockée le long de ces lignes, en utilisant un système de murs de soutènement, de remblais et de terrasses qui s'adaptent continuellement à la topographie environnante. Plutôt que d'imposer un paysage artificiel, cette approche amplifie la dynamique écologique existante du site, favorisant l'émergence d'un écosystème naturel autonome qui optimise la distribution de l'eau, favorise la croissance de la végétation et réduit les besoins d'entretien.

Les bords (edges) du parc sont conçus pour maximiser l'interface entre la ville et le paysage, en augmentant l'interaction entre la vie urbaine et les espaces verts tout en améliorant la valeur immobilière des développements adjacents. À l'image des limites changeantes d'un delta fluvial, où la terre et l'eau s'entremêlent dans une bordure en constante évolution, la frontière urbaine ondulée prolonge la relation physique et visuelle entre l'environnement bâti et la nature. Un système d'ondulations macro et micro à l'intérieur de cette bordure crée une inversion dynamique entre la ville et le parc, brouillant la frontière entre l'urbanité et la nature sauvage.

2.5 OPEN STACKS
Centre citoyen de Nantes

Nous avons tiré parti de la forme atypique du terrain sur lequel le projet est construit pour créer un bâtiment à la fois unique et véritable point de repère pour les espaces publics du quartier des Dervallières. Notre projet pour le nouveau centre administratif et social occupe la surface maximale du terrain, de manière à développer un bâtiment bas, proche du sol et facile d'accès, avec un espace d'accueil central pour le grand public aussi vaste que possible. En plan, son empreinte est en forme de pentagone irrégulier, évidé en son centre. L'entrée du bâtiment est en alignement avec l'axe de circulation du quartier pour faciliter sa lecture et son accès. Nous avons choisi de proposer un bâtiment avec une grande proportion de verre pour offrir au quartier un espace lumineux et accueillant.

Par la création d'un dénivelé permettant une lecture horizontale du bâtiment, et par une sérigraphie originale sur les vitres du premier étage, le bâtiment contribue à renouveler l'image du quartier. Avec la mise en œuvre de toitures végétalisées complètes ou partielles et du jardin suspendu intérieur, le projet s'intègre harmonieusement dans un site riche en verdure. C'est un bâtiment "topographique", qui fonctionne par entités programmatiques cohérentes qui permettent d'articuler les trois fonctions du bâtiment: Centre d'accueil social, centre de recherche d'emploi et centre d'information territorial

En façade, nous avons souhaité que chacun de ces niveaux soit identifiable par un système de façade qui lui soit propre. Le rez-de-chaussée s'ouvre sur la ville par de grandes baies vitrées, le niveau du premier étage par un mur rideau sur un parapet avec une sérigraphie évoquant les structures arborescentes du jardin que nous avons suspendu au-dessus du hall, et aux deuxième et troisième étages, un système de porte-à-faux et de dalles en surplomb protège les fenêtres du soleil et ouvre la cafétéria sur un jardin suspendu.

2.6 ACCUEILLIR LA BIODIVERSITÉ
Réhabilitation de l'Université de Bourges

Le campus de l'université d'Orléans à Bourges est composé en grande partie de bâtiments datant des années 1960, qui ont peu évolué depuis leur construction. Le défi de cette rénovation était de transformer ces structures vieillissantes en espaces d'apprentissage efficaces et contemporains tout en améliorant considérablement leur performance énergétique. Le projet comprend un bâtiment abritant deux amphithéâtres, un département de génie civil de 2 500 m^2 sur trois niveaux, une salle de travaux pratiques et un centre d'apprentissage de plain-pied, chacun étant conçu pour favoriser l'innovation dans les méthodes pédagogiques et le développement durable.

L'un des principaux objectifs de la rénovation était de repenser les salles de conférence traditionnelles. Comment un étudiant peut-il rester concentré pendant plusieurs heures dans un amphithéâtre de 300 places, alors que des études montrent que l'attention diminue généralement après seulement 20 minutes? Pour y remédier, la conception a introduit des amphithéâtres "en grappe", structurés pour soutenir à la fois les conférences et le travail de groupe. Chaque rangée de sièges en gradins est divisée en îlots de travail pour quatre à cinq étudiants, avec un banc incurvé et une tablette numérique commune. Cette configuration permet aux étudiants de collaborer en petits groupes tout en maintenant l'interaction avec le professeur. Développée en étroite collaboration avec le corps enseignant, cette disposition transforme l'amphithéâtre en un espace d'apprentissage flexible et participatif, pouvant accueillir à la fois des séances interactives et des cours magistraux traditionnels.

La rénovation a également intégré une approche innovante de la biodiversité, reconnaissant le potentiel écologique du campus. Lors de l'évaluation des bâtiments existants, une importante population de chauves-souris (environ 120 individus) a été découverte, nichant dans les fissures de l'isolation en polystyrène des façades. Plutôt que de les déplacer, le projet a adopté une stratégie de co-conception bio-inspirée, garantissant la présence continue de ces espèces tout en préservant l'intégrité de l'enveloppe du bâtiment.

Les façades, en revêtement en pierre composite sur isolation extérieure, intègrent des nichoirs à chauves-souris conçus sur mesure, parfaitement intégrés à la conception architecturale. Ces longues fentes de nidification sont dissimulées derrière le revêtement et réparties stratégiquement pour assurer à la fois l'accessibilité et la sécurité des chauves-souris. Pour éviter tout impact sur les utilisateurs, une tôle de zinc a été ajoutée sous chaque nichoir afin de protéger la façade et les zones environnantes des déjections.

Cette initiative écologique a été développée en collaboration avec les conservateurs du MHN de Bourges, experts internationalement reconnus dans la conservation des chauves-souris. Leur contribution a permis d'affiner les dimensions des nichoirs, leur orientation et les conditions microclimatiques, garantissant un habitat optimal tout en minimisant les perturbations potentielles pour les occupants du bâtiment. La conception privilégie la stabilité, la protection contre les prédateurs et une intégration discrète, ce qui en fait un modèle de préservation de la biodiversité urbaine.

Depuis l'inauguration du bâtiment, des colonies de chauves-souris ont colonisé les nichoirs, et des études écologiques ont enregistré une augmentation significative de la population locale de chauves-souris. En deux ans, le nombre de chauves-souris qui s'abritent à l'université est passé à environ 200, contre 110 avant la rénovation. Malgré leur présence, l'isolation extérieure reste intacte, ce qui démontre l'efficacité de la stratégie de conception pour concilier la conservation de la faune et la résilience architecturale. Ce projet représente un exemple pionnier de la manière dont la rénovation architecturale peut harmoniser pédagogie, durabilité et biodiversité, créant un campus à la fois innovant et écologiquement responsable.

MODÈLES DE SIMULATION POUR UNE ARCHITECTURE AUTONOME EN ÉNERGIE

Nous encourageons les architectes et les designers à intégrer des dispositifs de récupération d'énergie afin d'équilibrer la consommation énergétique des bâtiments. Cette approche vise à minimiser les coûts de maintenance et à répondre aux critères d'énergie zéro nette. Comment les méthodologies de conception architecturale évolueraient elles si une étude précise de l'efficacité énergétique globale pouvait cartographier les régimes de vent autour d'un bâtiment projeté. En réponse à cela, les réglementations urbaines pourraient-elles être adaptées pour optimiser le flux d'air en modifiant les profils des bâtiments?

Les recherches menées par notre équipe proposent une méthodologie de conception simple et efficace pour intégrer des dispositifs de captage de l'énergie éolienne dans les projets architecturaux. En analysant les flux d'air autour de différentes typologies architecturales, nous étudions l'impact des environnements urbains sur le comportement du vent. Cette recherche pourrait permettre d'identifier les emplacements optimaux pour micro-dispositifs éoliens en exploitant une aile accélératrice, appelée "aérofoil", pour améliorer l'extraction d'énergie. Cette méthodologie prend en compte l'orientation et l'échelle du bâtiment, les conditions contextuelles et les données éoliennes sur site pour optimiser les performances des dispositifs intégrés au bâtiment (BID), garantissant ainsi une utilisation efficace des ressources sur l'ensemble de l'enveloppe du bâtiment.

Analyse des données contextuelles aérauliques
La direction et la vitesse du vent ont une incidence significative sur l'énergie éolienne extractible. En milieu urbain, les bâtiments et les obstacles créent des flux d'air complexes. Des facteurs tels que la forme, la hauteur et l'espacement des bâtiments influencent la direction et l'intensité du flux éolien, ce qui a un impact sur la production potentielle d'énergie.

Dans les canyons urbains, le flux d'air est régi par la géométrie des rues et la répartition de la hauteur des bâtiments. La vitesse du vent est généralement plus élevée lorsqu'il souffle parallèlement à une rivière, mais diminue lorsqu'il est perpendiculaire en raison de la traînée aérodynamique. Des simulations de dynamique des fluides computationnelles (CFD) ont été combinées à des mesures du vent sur site pour établir une cartographie précise du vent urbain. Cette méthode a révélé l'impact profond des volumes et des obstacles urbains sur la vitesse et la direction du vent, soulignant la difficulté de prévoir les régimes de vent sans modélisation précise.

Au lieu de se fier uniquement à des analyses théoriques du vent, cette étude intègre des données provenant d'anémomètres connectés pour établir des cartes localisées des vents urbains. Ces études ont rapidement démontré le fort impact de la morphologie urbaine sur le flux d'air, montrant comment la vitesse et la direction du vent peuvent augmenter ou diminuer de manière significative autour des bâtiments. La variabilité du comportement du vent, influencée par les blocs urbains et les obstacles, rend difficile la prévision des modèles sans simulations précises. L'étude met en évidence la manière dont les configurations urbaines peuvent créer des zones de forte turbulence et des forces éoliennes multidirectionnelles au sein d'une empreinte urbaine compacte.

Simulation aéraulique sur un jumeau numérique urbain

Grâce à une pratique architecturale approfondie, nous avons développé un modèle 3D volumétrique d'un bloc urbain pour évaluer le comportement du vent sur les toits groupés dans les zones denses. Cette méthode permet:

– La catégorisation des différents types de vent en milieu urbain
– L'identification des vitesses de vent minimales et maximales
– La détermination des emplacements optimaux pour les dispositifs de récupération d'énergie

En intégrant l'analyse du vent sur site aux simulations CFD, nous avons mesuré l'augmentation de la vitesse du vent, qui influence directement l'énergie extractible (proportionnelle au cube de la vitesse du vent, selon le physicien allemand A. Betz). L'environnement virtuel de simulation permet une personnalisation poussée des paramètres, garantissant des simulations précises qui correspondent aux mesures réelles.

De nombreuses années d'expérience dans la simulation architecturale nous ont permis d'affiner notre approche, en créant des modèles qui représentent avec précision le comportement du vent en milieu urbain. Le modèle volumétrique tient compte des variations altimétriques, de l'orientation spatiale et de la densité des bâtiments, qui influencent tous l'accélération du vent et la turbulence. L'objectif est d'améliorer la production d'énergie en déterminant les meilleurs emplacements possibles pour les éoliennes dans un contexte urbain. En outre, la recherche étudie comment les phénomènes d'accélération du vent se produisent à des points de rupture spécifiques dans le tissu urbain, tels que les bords de toit et les angles de bâtiments, où la vitesse du vent peut être considérablement amplifiée.

Aérofoils et typologies d'intégration architecturale

Les "aérofoils", inspirés de l'ingénierie nautique, régulent et accélèrent la vitesse du vent aux abords des bâtiments, en exploitant le principe de Bernoulli et l'effet Venturi. Notre étude a permis d'identifier des éléments architecturaux clés, tels que les auvents, les marquises et les espaces interstitiels, qui concentrent naturellement le flux d'air, ce qui les rend idéaux pour la récupération d'énergie.

Les micro-éoliennes, classées en fonction de leur surface balayée (<25 m^2) et d'une puissance nominale allant jusqu'à 6 kW, peuvent tirer un avantage considérable de l'amplification du vent urbain. Dans nos conditions de simulation, la densité de l'énergie éolienne a été multipliée par un facteur de 3 à 8, tandis que la vitesse du vent a été amplifiée de 1,5 à 2 fois en raison des effets induits par les bâtiments.

L'étude a également analysé comment l'intégration de dispositifs d'accélération du vent, tels que les ailerons, peut améliorer la stabilité des flux d'air autour des bâtiments. En concentrant et en dirigeant les courants de vent, ces éléments atténuent les turbulences et améliorent l'efficacité des éoliennes. Les ailerons agissent comme des surfaces de contrôle qui modulent le flux d'air, permettant aux architectes de concevoir des structures qui contribuent activement à la production d'énergie. En outre, nous avons étudié comment les différentes typologies de BID influencent la production d'énergie, en tenant compte des variations de hauteur, de l'exposition au vent et de la densité urbaine.

Les ailes, initialement développées pour des applications maritimes et aéronautiques, se sont avérées efficaces dans des contextes architecturaux. En examinant le comportement du flux d'air le long des surfaces des bâtiments, nous avons identifié les zones où les dispositifs de récupération d'énergie peuvent être placés le plus efficacement. L'étude aborde également l'intégration d'ailes avec des turbines montées en façade, démontrant comment ces éléments peuvent stabiliser le flux d'air turbulent et augmenter l'efficacité de la capture du vent.

Intégration de dispositifs de récupération d'énergie dans l'enveloppe architecturale
La turbulence urbaine nécessite un placement stratégique des éoliennes pour assurer une production d'énergie stable. L'intégration des dispositifs BID dans les formes architecturales peut être classée soit comme des éoliennes autonomes (avec une installation et une maintenance simplifiées), soit comme des éoliennes rénovées (avec des performances variables en fonction du positionnement, de la stabilité structurelle et de la dissipation des vibrations).

Pour valider notre méthodologie, nous avons analysé l'école d'architecture ENSA Paris-Val de Seine, conçue par Frédéric Borel en 2008. Ce bâtiment a été choisi pour sa structure en forme de bloc, son emplacement stratégique dans une zone à fort vent et son accessibilité pour la surveillance de la vitesse du vent. Les simulations CFD ont montré une amplification significative du vent entre ses deux tours, avec des vitesses passant de 5 m/s à environ 17 m/s.

Pour exploiter cet effet, nous avons proposé des panneaux aérodynamiques rigides pour concentrer et stabiliser le flux du vent. En ajustant les angles des profils aérodynamiques, les vitesses du vent entre les tours ont été optimisées à 16-18 m/s, ce qui a permis de compenser environ 18 % de la consommation d'énergie annuelle du bâtiment grâce à l'intégration des BID.

En outre, l'étude explore les stratégies potentielles d'optimisation de la production d'énergie éolienne urbaine en envisageant de nouvelles approches de conception des façades. Plutôt que de traiter les surfaces des bâtiments comme des éléments passifs, cette recherche prône une intégration active, transformant l'architecture urbaine en un écosystème générateur d'énergie. L'utilisation de techniques d'amplification du vent combinées à des éléments BID permet une utilisation plus efficace des ressources, contribuant à la durabilité à long terme dans les environnements à forte densité.

Enfin, nos recherches combinent des simulations CFD avec des mesures du vent en temps réel pour affiner les évaluations du potentiel éolien en milieu urbain. En identifiant les zones d'accélération du vent localisées, nous fournissons une méthodologie pour optimiser le placement des éoliennes à axe vertical, améliorant ainsi leur efficacité dans les environnements urbains denses. Cette approche permet aux architectes d'intégrer de manière transparente des solutions de captage du vent dans la conception des bâtiments, faisant ainsi progresser l'architecture durable vers l'autonomie énergétique.

L'intégration de profils aérodynamiques et de wdispositifs BID représente une avancée significative dans la redéfinition des formes architecturales visant à optimiser l'utilisation de l'énergie éolienne. En exploitant les couloirs de vent urbains et en optimisant les profils des bâtiments, nous proposons un nouveau paradigme de conception qui donne la priorité à l'efficacité énergétique. La méthodologie décrite dans cette étude offre aux architectes un cadre stratégique pour concevoir des bâtiments qui non seulement consomment moins d'énergie, mais contribuent également à la production d'énergie renouvelable, conformément aux objectifs contemporains de développement durable.

Les recherches futures se concentreront sur l'amélioration des techniques d'analyse du vent, le développement de systèmes de façades adaptatives et l'exploration de nouveaux matériaux qui améliorent l'efficacité énergétique. En combinant l'innovation architecturale avec des technologies de simulation avancées, cette étude jette les bases de la prochaine génération de bâtiments à énergie positive.

3. MATTER
DE LA MATIÈRE INFORMÉE À LA MATIÈRE RÉEMPLOYÉE

Matériaux du passé au présent

La notion de matière est explorée dans le projet en tant que dispositif architectural, pour créer un dialogue avec l'histoire du lieu dans certains projets, ou pour créer une expérience physique ou conceptuelle de la matière et dans d'autres, ou encore comme mode constructif alternatif en matériau de remploi dans une posture militante de préservation des ressources naturelles.

Dans de nombreux projets la matière constitue un mode de dialogue puissant avec le passé du lieu. Le choix d'un matériau puisé dans les ressources locales et historiques permet de tisser des liens et d'ouvrir une relation avec des bâtiments adjacents ou patrimoniaux, soit par similarité, soit par contraste avec une condition existante. Cette stratégie de dialogue est mise en œuvre dans le projet pour la bibliothèque de Saint-Hilaire-du-Harcouët, pour le musée de la bataille de Fromelles et pour la médiathèque de Lezoux.

A Saint-Hilaire, une recherche acharnée de pierre dans les carrières de granite local a permis de remonter à l'origine du granite rose singulier avec des taches de rouille qui est mis en œuvre au XIIIème siècle dans l'église classée en face du bâtiment.

Pour le musée de la bataille de Fromelles, c'est l'évocation de stratifications géologiques des terres des fouilles archéologiques mais aussi des tranchées de la première guerre mondiale qui est enchâssé dans les murs du musée par un béton de terre coulée par strates successives.

Dans la médiathèque de Lezoux, la présence d'un site majeur de fabrication de poteries romaine sigillé du Vème siècle avant notre ère qui a orienté l'utilisation de la briques cuite pour les murs du bâtiment, avec la réécriture de motifs d'empilement, à la fois évocation de livres sur des étagères, systèmes d'ouies pour des ventilation du bâtiments, ou encore perforation acoustique pour le traitement des espaces de travail de la médiathèque.

Dans les projets du Glass Loft à New York, et des réserves des musées de Bayonne, la matière est avant tout un dispositif conceptuel qui propose une expérience globale de l'architecture, une atmosphère dont le projet se fait le lieu d'une expérience physique et personnelle. Le matériau évoque par l'expérience un rapport au lieu, au corps et au site.

Dans le cas des Réserves de Bayonne, le volume entièrement opaque des réserves pour les œuvres d'art des musées du pays basque génère l'opportunité de la généralisation d'une enveloppe-paysage. Constitué d'une peau acier inox poli miroir martelée, le bâtiment réfléchit par touches le paysage des hauts de Bayonne, à la manière d'une peinture impressionniste de Claude Monet, transformant le reflet de la ville et du ciel en une expérience atmosphérique de la couleur et du paysage.

Le glass loft est avant tout une exploration de la qualité cristalline du verre pour qualifier la relation entre espace intime et espace de vie de ce loft NewYorkais. Le verre n'est pas utilisé pour sa transparence, mais plutôt pour sa qualité structurale et sculpturale afin de définir l'"architecture intérieure d'un espace ouvert d'habitation.

L'atelier des arbres aborde la question de la matière par le matériau de construction de réemploi. Le projet est une tentative vertueuse et militante de réutiliser des structures et matériaux issues d'autres chantiers ou de la démolition de bâtiments remplacés pour créer un espace de travail singulier en cœur d'îlot-jardin. Les arbres existants sur le site ont été incorporés au projet pour qualifier les espaces intérieurs de l'atelier et favoriser l'émergence d'un écosystème vivant en symbiose avec les utilisateurs de l'espace.

Activer une économie circulaire des rossources matérielles

Cet essai, issu d'une recherche avec D. Toubanos et D. Sipp, interroge la transformation des pratiques architecturales et urbaines dans un contexte de raréfaction des ressources matérielles et de réflexion globale sur la durabilité des bâtiments et leur cycle de vie. Cette crise des ressources primaires oblige les architectes à développer des méthodes alternatives de conception et de construction du projet en partant du principe que la ressource matérielle est à l'origine de la genèse du projet, en amont du concept architectural.

Le travail commence par l'identification et la redécouverte des sources des matériaux et des produits disponibles pour l'architecture. Le choix des matériaux, parfois laissés aux mains des constructeurs, redevient un élément fondamental du projet pour les architectes qui définissent leurs spécificités ou leur origine, en relation avec leur impact sur le cycle de vie global du bâtiment.

Par une expérience pédagogique, cette démarche vise à explorer la collecte et le référencement de ces gisements. La géolocalisation de ces ressources par une technologie de traceurs GPS, permet le référencement et la réutilisation en continu de matériaux inutilisés. C'est la voie du réemploi en architecture. Les matériaux issus de la déconstruction, les produits de la production industrielle ou artisanale, les invendus ou les retours clients, la ville, ses entreprises et ses usagers regorgent de ressources potentielles que les architectes doivent explorer dans leur production. L'objet trouvé, le détournement, la réutilisation, le recyclage sont autant de stratégies qui participent à la créativité et à l'invention du projet, ainsi qu'au développement d'une matérialité et d'une image alternative du projet. Cet essai évoque aussi comment prototyper des structures et les tester pour évaluer les performances de matériaux et d'assemblages inconnus? Il s'agit d'une mission complexe où la méthode scientifique offre un cadre et une approche pour étudier et découvrir ces nouveaux potentiels.

Des territoires physiques et virtuels ouverts

Nous évoquons ici la migration d'une partie de nos environnements physiques vers un environnement virtuel est en cours et poussera progressivement les architectes à intervenir, comme les métropoles, les environnements ruraux, les réseaux sociaux.

Concevoir l'architecture, c'est puiser dans nos expériences émotionnelles les plus personnelles, mais que deviendra cette architecture lorsque les architectes auront passé plus de temps dans des mondes immersifs que dans des espaces physiques?

Dans la première réalité, la plupart de nos expériences et émotions sont également générées par les relations sociales, par nos rencontres ou interactions. C'est pourquoi nous travaillons sur un modèle de conception en réalité augmentée, et non sur celui d'un monde virtuel fermé. Dans ces mondes virtuels fermés, en opposition à notre réalité physique, nous ne serons jamais surpris par un chant d'oiseau inattendu dans cet environnement, un brouhaha qui révèle la complexité de l'occupation des espaces, par la fraîcheur de l'air dans un espace extérieur, l'odeur d'une mousse en traversant un espace, la chaleur d'un rayon de soleil ou un souffle sur notre visage. Imaginez un monde où les simples 3 dimensions, 4 ou 5, sont remplacées par une réalité hybride complète, une réalité multidimensionnelle dans le temps et l'espace, incluant tous les utilisateurs.

Pour nous, c'est un monde hybride complet où la réalité primaire et toutes ces composantes physiques, climatiques, biologiques, chimiques se combinent et amplifient leur expérience les unes avec les autres. Construire avec des matériaux hétérogènes et variables, des bio-matériaux, associés à des matériaux industriels permettant une hybridité de compétence et de matérialité. Face à la beauté et à la complexité de notre réalité, la création de modèle en réalité augmentée n'est qu'une étape intermédiaire dans l'expérience d'une réalité recomposée par l'homme.

3.1 LES PIERRES RETROUVÉES
Médiathèque de Saint-Hilaire

La nouvelle médiathèque de Saint-Hilaire-Du-Harcouët est située dans un contexte historique et urbain exceptionnel, face à une église en pierre datant du XIIIe siècle. Conçu pour prolonger le tissu urbain environnant, le bâtiment est volontairement bas pour respecter son contexte historique et est partiellement encastré dans le sol. Cette récession accentue la terrasse inférieure, rendant la bibliothèque visible depuis le parc.

Des matériaux tels que le granit bleu caractéristique, avec ses taches de rouille naturelle, font écho à ceux utilisés dans la construction de l'église au XIIIe siècle. Les fenêtres allongées reflètent les proportions et l'évasement uniques des fenêtres de l'église elle-même, un bâtiment public qui a servi de référence clé pour notre conception. Sur le côté sud, une fenêtre allongée donne sur le lac de la ville, offrant une vue sur l'église et ses structures voisines.

La médiathèque est conçue comme un sanctuaire où la lumière naturelle joue un rôle essentiel, créant un environnement propice aux échanges, un lieu où le savoir est accessible à travers différentes disciplines et cultures. Les vastes surfaces vitrées du rez-de-chaussée ouvrent davantage le projet sur la ville.

À l'intérieur, les espaces de la bibliothèque sont enveloppés par une façade à double vitrage, dont une couche est gravée d'un treillis métallique dont le motif s'inspire de la rosace de l'église. Cette conception crée une lumière douce et diffuse, idéale pour lire et travailler. De plus, notre approche exploite la profondeur de la façade, en utilisant la distance entre les brise-soleil et le mur vitré ainsi que l'espacement des lames, pour créer des projections et des surplombs dynamiques. Il en résulte une série d'espaces riches et transitionnels qui estompent la frontière entre l'intérieur et l'extérieur.

3.2 MUSÉE DE LA BATAILLE DE FROMELLES

Le 19 juillet 1916, plusieurs bataillons de la 5e division australienne passent vaillamment les lignes du front allemand. Malheureusement, les allemands les encerclent définitivement jusqu'à la fin de la bataille en matinée du 20 juillet. Les blessés seront soignés, les prisonniers interrogés, les morts rassemblés et enterrés dans cinq fosses communes qui resteront ignorées jusqu'à ce qu'un historien local signale en 1999 ce lieu d'enfouissement. Suite à une recherche de terrain puis une exhumation en 2009 commanditée par la Commonwealth War Graves Commission pour le compte des gouvernements australiens et britanniques, 250 corps ont été retrouvés et un cimetière leur a été créé sur le territoire de Fromelles pour les accueillir dignement. C'est sur ce territoire que se trouve le musée consacré à cette bataille.

L'architecture du musée de Fromelles est conçue pour créer un lien entre le ciel et la terre, unissant les visiteurs au lieu de repos solennel des soldats. La manipulation sensible de la topographie du site accentue sa pente naturelle, enveloppant partiellement le musée tout en permettant à la salle d'exposition principale de sortir d'un monticule de végétation. Cette émergence, réalisée en béton brut par coulées successives de béton de tonalité différentes, évoque les couches géologiques du sous-sol, celui des tranchées, mais aussi de l'enfouissement des corps. Le musée se révèle progressivement aux visiteurs, grâce à un profil bas qui répond au cimetière militaire adjacent.

Deux axes visuels clés déterminent l'orientation du musée. L'un mène vers le cimetière, reliant le dernier segment de l'exposition au lieu de repos des soldats, tandis que l'autre ouvre le hall vers l'église de Fromelles, une structure proéminente qui définit sans équivoque la ligne d'horizon de la ville. Ces axes aident à orienter et à articuler les facettes de l'espace d'exposition octogonal.

Au cœur du musée se trouve un noyau central octogonal qui forme la salle d'exposition principale. Cette conception, en résonance avec le cimetière adjacent, rappelle à la fois la géométrie rigide des structures militaires et la pureté sacrée illustrée par le tambour octogonal de l'église de Fromelles.

La lumière naturelle joue un rôle essentiel dans tout le musée. Les espaces d'accueil du public sont baignés de lumière naturelle grâce à des puits de lumière stratégiquement placés le long des murs.

L'espace d'exposition commence dans l'obscurité totale et est progressivement éclairé par des puits similaires le long de son périmètre. Le soir, l'éclairage extérieur définit subtilement les espaces commémoratifs autour des champs de bataille, offrant une scénographie lumineuse qui renforce le caractère réfléchi du bâtiment.

La muséographie se concentre principalement sur la bataille de Fromelles tout en la contextualisant dans des récits internationaux, nationaux et locaux. Dès l'entrée, les visiteurs empruntent une passerelle surélevée, évocatrice de fouilles archéologiques, qui les guide à travers une série d'expositions. Flanqué de vitrines, ce chemin se transforme en tranchée symbolique. La visite se termine dans la "War room", un espace circulaire rappelant un centre de commandement, où les événements de la bataille de Fromelles sont présentés par une installation vidéo sur une table centrale.

Un lien entre le ciel et la terre

L'architecture du musée de Fromelles cherche à établir un lien entre le ciel et la terre, entre les visiteurs et "l'enfouissement" des soldats. Un travail sur la topographie a permis d'accentuer la pente naturelle du site et de couvrir en partie la zone technique du musée.

Seule la salle d'exposition émerge de cette butte végétale, en indiquant la trace d'un enfouissement certain. Cette émergence est en béton brut, dont les coulages successifs évoquent l'empilement des couches géologiques et des natures de terre différente du sous-sol.

Le musée se dévoile ainsi progressivement aux visiteurs qui arrivent sur le site, par son profil bas, qui est respectueux de la présence du cimetière adjacent.

Deux axes visuels ont généré l'implantation du musée. Un axe vers le cimetière qui permet de mettre en relation la dernière séquence de l'exposition avec le lieu de repos des soldats. Un axe visuel ouvre le hall vers l'église de Fromelles, édifice élevé qui identifie immédiatement la silhouette de la ville. Ces axes ont permis d'orienter et de définir les «facettes» de l'espace d'exposition de forme octogonale.

Un noyau central octogonal évoquant les constructions militaires et sacrées

Le musée est organisé autour d'un noyau central de forme octogonale qui constitue la salle d'exposition principale. Cette forme, dans l'esprit du cimetière adjacent, évoque à la fois la géométrie rigide des constructions militaires, ainsi que la géométrie pure du sacré tel le tambour octogonal de l'église de Fromelles.

La toiture met en scène la division du projet entre le volume de la salle d'exposition et celui des locaux d'administration. Un dispositif de faille de lumière zénithale entre ces deux éléments permet de faire largement pénétrer la lumière dans le hall. La bande à l'Ouest regroupe les espaces administratifs et techniques du musée tels que l'administration, l'accueil, le vestiaire, les sanitaires et l'espace cafétéria.

Ambiances intérieures

La lumière naturelle joue un rôle important dans les espaces du musée, elle est au coeur du projet. Les espaces d'accueil du public sont largement baignés de lumière naturelle, par une faille lumineuse qui longe les murs. L'espace d'exposition démarre dans l'ombre totale et progressivement s'illumine par un dispositif similaire de faille lumineuse en périphérie des espaces.

En fin de journée, des lampes installées dans ces "failles" et reliées à des cellules photosensibles compenseront progressivement le manque de lumière naturelle.

Dans la soirée, les espaces extérieurs du musée sont balisés par des diodes électroluminescentes à très longue durée de vie et consommation d'énergie réduite. Ce balisage permet un repérage nocturne subtil des espaces de mémoire des champs de bataille alentours et propose véritablement une scénographie lumineuse du bâtiment.

Une scénographie centrée sur la mémoire

En contact visuel direct avec le cimetière, le parcours de l'exposition rappelle par sa géométrie le plan à facettes du cimetière. La muséographie est essentiellement dédiée à la bataille de Fromelles mais en la reliant à son contexte international, national et local. Le flux de visiteur doit tenir compte de l'exiguïté du bâtiment, de la composition humaine et sociale des visiteurs attendus et des règles en matière de sécurité et d'handicap.

Dès l'entrée, le visiteur emprunte une passerelle "traversant" les différentes séquences. Elle est suspendue au dessus du sol. C'est une référence aux fouilles archéologiques. Sa qualité est de canaliser le parcours en "accompagnant" didactiquement le visiteur. Des sons et divers bruits sont diffusés sur le parcours. Ils le sont par des hauts parleurs directionnels situés opportunément au long du parcours.

De chaque côté, il avance ensuite sur la passerelle qui devient tranchée, entre les vitrines qui présentent les collections.

Le point central de la visite est la "War Room", espace circulaire, telle une salle de commandement dans laquelle sera présenté, par une projection vidéo sur une table basse, le déroulement de la bataille de Fromelles.

Des alcôves, dilatations le long du parcours, permettent ensuite au visiteur de découvrir des parcours de vie de soldats, dans des espaces intimes propres au recueillement.

3.3 LIRE DANS UN JARDIN

Médiathèque de Lezoux

La nouvelle médiathèque de Lezoux occupe un site qui offre un espace de respiration au sein de la ville, avec une vue dégagée sur le paysage environnant. Pour répondre à ce cadre unique, nous avons proposé une structure de type pavillon, un lieu d'accueil où livres, ressources multimédias et activités diverses sont accessibles au milieu d'un jardin luxuriant. Son design transparent estompe les limites du site, transformant le bâtiment en véritable vitrine urbaine.

La structure est dotée d'un toit ajouré qui abrite un volume transparent abritant les espaces de consultation et la salle de lecture. Soutenu par des piliers en acier, le toit s'appuie sur deux masses de béton opaques qui servent d'éléments de contreventement. Sa silhouette élancée et son design perforé évoquent l'ambiance lumineuse du sous-bois. Des stores intégrés renforcent la protection solaire et le confort visuel, créant un environnement propice à la lecture et au travail.

À l'intérieur, la médiathèque est conçue comme un espace polyvalent aux frontières fluides, favorisant l'interaction entre toutes les générations. L'espace respire la transparence et facilite l'accès à ses collections, passant en douceur d'un grand espace d'accueil à des zones plus calmes, idéales pour la concentration et le travail individuel, et menant progressivement à des espaces plus intimes, réservés au personnel.

Deux salles "poreuses", situées en face de la salle de lecture, offrent des espaces de travail calmes et intimes qui permettent aux utilisateurs de s'isoler complètement de leur environnement ou, au contraire, de se connecter avec l'extérieur via la terrasse de lecture adjacente.

Autour du bâtiment, une série de jardins distincts animent les espaces intérieurs de la médiathèque. Des bouleaux (Betula papyrifera) ont été soigneusement plantés sur l'ensemble du site pour accentuer ces espaces extérieurs. L'"Arbre de la Sagesse", avec son feuillage lumineux et son écorce blanche semblable à du parchemin, fait notamment écho aux colonnes blanches extérieures du bâtiment.

L'accès à l'avant-cour se fait par un dallage en pierre de lave andésitique, tandis que le terrain environnant est sculpté en douces buttes recouvertes d'une prairie diversifiée. Ces buttes soutiennent un théâtre en plein air qui s'ouvre sur la scène extérieure de la salle polyvalente du côté ouest, complétée par des terrasses en bois qui prolongent les activités de la bibliothèque à l'extérieur.

3.4 GLASS LOFT

New York

La conception de ce mur de verre pour un loft résidentiel à New York s'est inspirée de la nature cristalline du matériau verre.

Le mur oscille à l'intérieur du loft, délimitant les espaces de vie, les chambres et les zones de service tout en régulant à la fois les vues et l'accès entre eux.

Sa structure s'entrelace avec la charpente en bois existante du bâtiment, parfois ancrée à celle-ci, parfois en porte-à-faux au-delà, créant un espace d'ambiguïté intense entre une structure rationnelle et une enveloppe de forme libre. La transformation de la géométrie du mur a été influencée par le mouvement du corps humain lorsqu'il traverse l'espace et interagit avec ses limites.

En infléchissant le vecteur par lequel le corps traverse le mur, la surface gagne en épaisseur et se dédouble effectivement, offrant ainsi l'intimité nécessaire aux deux chambres situées derrière.

3.5 LES RÉSERVES DES MUSÉES DU PAYS BASQUE

Bayonne

Le projet de construction de réserves muséales mutualisées de la ville de Bayonne s'inscrit dans la volonté de la ville d'assurer la préservation de son patrimoine artistique et culturel. Les réserves permettent de contrôler précisément les conditions de conservation de chacune de ces œuvres d'art en adéquation avec leurs typologies, leur origine ou leur état actuel.

Le projet propose la division des zones de stockage en plus petites cellules , augmentant la polyvalence dans l'utilisation de ces espaces et dans la répartition des objets dans les salles de stockage. C'est un bâtiment de forme "minimaliste", où l'organisation interne, basée sur l'empilement d'espaces de stockage, génère le plan du bâtiment.

Le projet des Réserves des musées de Bayonne est issu d'un travail sur la matérialité de l'architecture pour évoquer les notions de permanence (celle des objets) et la notion d'incertitude (celle de l'instant). La façade est composée d'un bardage de larges panneaux d'acier poli miroir fabriqués par emboutissage numérique de feuilles d'acier inoxdable.

La forme des plis de la tôle évoque le travail artisanal du métal tel que le martelage des objets métalliques, ou l'aspect de surface de sculptures en bronze. Cette technique permet de rigidifier les panneaux de façade. Elle crée des motifs de forme libre sur la façade et avec une continuité parfaite des panneaux entre eux.

Ce dispositif, à l'échelle du contexte, se traduit par une véritable expérience de la lumière et des alentours du site. Comme dans certaines œuvres de peintres impressionnistes tels que Monet, les réflexions de lumière sur des surfaces libres permettent de révéler et de mettre en scène une expérience instantanée. Plutôt qu'un simple stockage d'œuvres d'art, les réserves sont conçues comme des témoins actifs du passage du temps, capturant chaque seconde avec intensité, grâce à une enveloppe architecturale dont la matérialité et la géométrie résument de manière unique ces moments transitoires.

3.6 L'ATELIER DES ARBRES

Paris-Cachan

Implanté dans la cour d'un immeuble résidentiel à Cachan, le projet prend vie au cœur d'un jardin arboré. Cet atelier ne pourrait exister sans préserver les arbres ayant poussé spontanément dans ce lieu. Nous avons fait le choix de conserver l'intégralité des arbres existants et de concevoir un bâtiment qui s'inscrit délicatement entre eux, entièrement réalisé en bois et en briques de réemploi.

Les ossatures bois de la toiture, conçues avec précision, se glissent entre les branches des arbres sans jamais les toucher, afin de respecter pleinement leur port naturel. Porté par un unique poteau en lamellé-collé, le bâtiment se limite à un rez-de-chaussée. Sa structure en bois est surmontée d'une charpente dotée d'un toit en shed avec une verrière orientée au nord, maximisant ainsi la lumière naturelle. Les arbres et le jardin environnant jouent un rôle central dans la régulation climatique du bâtiment, en offrant ombrage et fraîcheur grâce à leur feuillage.

Ce laboratoire de conception et d'expérimentation s'ouvre sur le jardin, transformé en une extension extérieure du bâtiment. L'architecture, par sa porosité aux éléments naturels et à la biodiversité, crée ce que nous appelons un "territoire partagé du vivant". C'est un espace où cohabitent et dialoguent la diversité du vivant: fourmis et pies bavardes, cerisiers et citronniers, fougères et orchidées, escargots et araignées. Ici, l'espace bâti ne s'isole pas de son environnement naturel et vivant; il devient un lieu de réconciliation entre les espèces végétales, animales et humaines.

Cette architecture propose une expérience unique: celle d'un espace vivant, atmosphérique et poreux. Parfois, le vent animant les arbres crée des crissements de la rappelant, évoquant le bruit d'une coque de bateau dans la mer. La pluie, guidée par le mouvement des branches, s'écoule directement le long des troncs des arbres, jusqu'à leur racine, renforçant l'intimité entre l'édifice et le jardin.

L'atelier des arbres est un lieu où le dedans et le dehors s'entremêlent, où les frontières disparaissent, un espace offrant une nouvelle manière d'habiter et de ressentir notre rapport au monde naturel.

L'IA GÉNÉRATIVE POUR LA CONCEPTION ARCHITECTURALE EN RÉEMPLOI MÉTHODOLOGIE ET EXPÉRIMENTATION À L' ÉCHELLE 1

Cette recherche présente une méthodologie alternative de conception architecturale en matériaux de construction de réemploi, en assemblant un gisement de matériaux identifiés à l'aide d'une IA générative de type texte2image. Cette méthode vise à dénombrer un jeu de ressources matérielles hétérogènes et un modèle de recommandation d'assemblage.

1. Contexte de la recherche

L'impact cumulé de l'activité de la construction est estimé de 45 à 55% des déchets produits dans le monde[1]. La responsabilité des architectes, des ingénieurs, comme des maîtres d'ouvrage, est incontestable, mais la construction de notre environnement et sa transformation offrent aussi un potentiel pour des solutions nouvelles. Elle ouvre une réflexion pour la transformation de nos modes de vie et de notre impact sur la terre. L'action d'habiter notre territoire ne doit plus se traduire par la construction de nouvelles "surfaces artificielles"[2] en remplacement de l'écosystème naturel. La pratique de l'architecture, à terme, ne devrait plus être de construire par addition et extraction de ressources, mais plutôt en inversant le cycle de la matière par la déconstruction du bâti et la reconstitution de nos ressources naturelles. Cette vision inversée de l'activité de l'homme est aujourd'hui à son point de basculement. La convergence des analyses de l'accélération du changement climatique et leur impact sur les écosystèmes vivants est à maturité et nous engage à déployer tous nos efforts dans l'accélération de ce changement de paradigme.

Cette recherche interroge la transformation des pratiques architecturales et urbaines dans un contexte de raréfaction des ressources matérielles et de la réflexion globale sur la durabilité des bâtiments et leur cycle de vie. Cette crise des ressources primaires astreint les architectes à développer des méthodes alternatives de conception et de construction du projet en partant du postulat que la ressource matérielle est à l'origine de la genèse du projet, en amont du concept architectural.

Cette approche du projet par le matériau oblige à revoir la notion de conception, ou plutôt de "Concevoir la conception" comme l'évoque le sociologue Edgar Morin, dans le tome 3 de son œuvre quasi-encyclopédique La Méthode[3] et qui suggère de prendre une distance avec l'objectif. Penser la méthode de l'architecture, c'est penser la conception plutôt que la simple finalité elle-même: l'architecture.

La question de la recherche en "design" mobilisent un cadre épistémologique spécifique et différent du cadre classique de la recherche scientifique, comme le décrit H. Simons[4], c'est-à-dire mobilier des savoir-faire pour la production d'artefacts, plutôt qu'un corpus traditionnel de recherche. C'est un domaine où la méthode scientifique offre un cadre de travail et une approche pour étudier et ouvrir de nouveaux potentiels de "design".

2. Méthodologie: Une IA générative pour la réutilisation innovante de matériaux de réemploi

Ce travail de transformation et d'inversion des pratiques de l'architecte démarre par l'identification et la redécouverte des gisements de matériaux locaux et produits disponibles et pour la construction. Souvent à la main des constructeurs ou des économistes de la construction, le choix des matériaux, leurs spécificités ou leur provenance, en relation à leur impact sur le cycle de vie global du bâtiment redevient pour les architectes un élément fondamental du projet.

Matériaux de déconstruction, produits de rebuts de production industrielle ou artisanale, invendus ou retours clients, la ville, ces entreprises et utilisateurs regorgent d'un potentiel de ressources que les architectes doivent explorer dans leur production. L'objet trouvé, le détournement, le réemploi, le recyclage sont autant de stratégies qui participent à la créativité et à l'invention du projet, aussi bien qu'au développement d'une matérialité et d'une image alternative du projet. En réponse à un projet de construction, nous avons donc lancé dans le cadre d'un exercice pédagogique en Master 1 à l'ENSAPVS, une expérience de découverte de ce gisement dans un cercle de 1 km autour de l'ENSAPVS.

Dans un deuxième temps, la méthode développée dans cette recherche met en œuvre des modèles d'IA génératifs créatifs, de type Text2Image, telles que les outils Midjourney, Dall.e, Gemini ou Crayon, pour proposer et inspirer une conception d'un projet à partir d'un ensemble de matériaux trouvés et disponibles pour le réemploi. Ces modèles génératifs ont été détournés de leur usage de simple illustration graphique pour mettre au point des propositions de design avec des matériaux de réemploi et une intention architecturale spécifique. Ces outils sont interrogés par des "prompts", un codage en langage naturel qui permet de décrire et de questionner un réseau de neurones entraîné sur un jeu de données graphiques et syntaxiques .

Le troisième aspect de la recherche est la mise en situation de ces techniques dans un chantier de construction à l'échelle 1 avec un groupe d'étudiants de master en école d'architecture. Nous prenons ici l'exemple du pavillon Archifolles, construit par les étudiants de l'ENSAPVS en Mai 2024, dans le cadre d'un projet financé par le Ministère de la Culture pour proposer une olympiade culturelle de mise en avant de l'architecture dans le cadre des Jeux Olympiques 2024 à Paris .

3. Le réemploi de matériaux en architecture

Les principaux inconvénients de la réutilisation des matériaux de réemploi qui ont été identifiés par les étudiants sont les suivants:

a. La tâche difficile d'identifier et de repérer les matériaux provenant des chantiers de construction, des sites de démolition et des déchets industriels.
b. Conception architecturale dans le cadre d'un stock de matériaux limité et hétérogène
c. Élaboration de scénarios de détails architecturaux et d'assemblages structurels

La méthodologie proposée est basée sur un processus qui utilise la capacité spécifique de l'intelligence artificielle générative texte2mage pour compter la quantité d'articles matériels et de suggestion d'assemblage selon des critères spécifiques.

Nous avons analysé 5 packages GenAI différents pour caractériser leur performance à compter et assembler un ensemble de matériaux donnés. L'utilisation de l'IA générative, sur une liste limitée et non prévisible de ressources matérielles, analyse la quantité, la forme, les caractéristiques physiques qui sont déconnectées de l'idée architecturale préconçue.

L'objet de l'expérimentation est d'évaluer les capacités des outils de conception numériques et des IA génératives à optimiser le processus de conception d'architecture en matériaux de réemploi. Ainsi il est primordial d'identifier ces capacités selon les critères qui sont inhérents au réemploi et à la construction de manière générale. Ils pourraient être: les matériaux, les échelles, les quantités, les dimensions, l'optimisation et l'économie, les assemblages et l'organisation ainsi que les modèles et références.

4. L'ingénierie du prompt pour l'idéation architecturale

Afin de générer des propositions pertinentes, la création de textes ou "prompt" a mis l'accent sur l'identification de variables claires mettant en évidence les critères relatifs à la construction en matériaux de réemploi. Les "prompts" les plus simples possibles et en anglais, dans les expériences, semblent éviter les éventuels biais par le recours à des références susceptibles de changer les résultats escomptés.

Nous ne pouvons détailler ici toutes les générations réalisées sur ce projet, mais nous donnons ici 4 exemples générés sur 2 plateformes génératives Midjourney et Dall.E.

Dans cet exemple, les images générées présentent le matériau utilisé dans une démarche innovante, économique et durable. La variable "réemploi" ou "reclaimed" en anglais est ensuite intégrée au «prompt» afin d'estimer la capacité des IA à repenser l'expression et la mise en œuvre du matériau. Ces images ont été générées par Tim Houplain et David Serero dans le cadre du séminaire Données et Territoires à l'ENSAPVS.
(Voir Figure 3, page72)

En comparant ces résultats, nous pouvons établir que la variable «réemploi» semble bien intégrée par les IA. Dall-E a une tendance à créer des projets monumentaux par leur volumétrie, s'affranchissant du critère économique.

Dans cette 2ème expérience, nous avons essayé d'organiser des éléments hétérogènes et d'apporter des solutions d'assemblages innovantes et inédites.
(Voir Figure 4, page 272)

Dans cet exemple, l'apparence des fenêtres est hétérogène. Leur assemblage suit une trame verticale avec des hauteurs variables mais leurs largeurs sont courantes. L'image proposée par Dall-E positionne des volets perpendiculairement à la façade, une analogie peut être faite avec les raidisseurs en verres parfois utilisés dans la mise en œuvre de murs rideaux ou "curtain wall", terme utilisé dans le "prompt".

Ces 2 exemples nous permettent de confirmer que ces plateformes d'IA génératives sont capables de proposer des organisations d'éléments hétérogènes avec une logique d'assemblage et de structuration identifiables. La mise en œuvre proposée par ces modèles pose des questions qui induisent des réponses parfois non pertinentes mais souvent propres à inspirer de vrais propositions de conception. Les étudiants ont expérimenté cette méthode tout au long de leur processus de conception.

Cette expérimentation nous permet de tenter de vérifier la capacité de ces modèles à quantifier dénombre, ou donner une ordre de grandeur des quantités de matériaux ou des échelles. Les premiers modèles de langages de ce type avaient parfois des difficultés à identifier certaines quantité simples (nombre de doigts, ou nombres de fenêtres, par exemple). Dans les requêtes posées dans le cadre architectural, les quantités sont souvent importantes et c'est véritablement l'échelle qui est considéré, plutôt que la quantité exacte.
(Voir Figure 5, page 274)

Suite à de nombreuses expériences de ce type, des échanges avec les futurs utilisateurs du pavillon et les contraintes techniques de réalisation du projet, les étudiants ont pu finaliser un projet. L'article ne présente pas ici le détail des étapes de la conception hybride (conception classique et conception augmentée) , mais présente quelques étapes de la construction.

Ici, les matériaux de réemploi ont été utilisés aussi bien pour les structures primaires que secondaires. L'étude de cas de quelques exemples qui mettent en œuvre des matériaux de réutilisation de nos jours montrent, que dans ces projets, les matériaux de réemploi sont principalement utilisés dans les structures secondaires, dans le cadre de «façade décorative» ou dans le cadre de parties non structurelles d'un bâtiment, telles que les murs de séparation, les portes, les fenêtres, etc.
(Voir Figures 7-8, page 276)

5. Construire une architecture à léchelle 1 en bois de réemploi

La dernière étape a été la réalisation à l'échelle 1 du pavillon sur le site du village Olympique du parc de la Villette qui a commencé avec des parties de l'ensemble du projet en faisant appel à la construction hors site, jusqu'à la construction sur site. Cela a permis de développer des compétences collaboratives des étudiants, afin d'unir leurs forces améliorant ainsi leur synchronisation intellectuelle et manuelle. Une partie importante de ce processus était la confrontation immédiate à la réalité.
(Voir Figures 9-10, page 276)

Bien que nous soyons conscients que l'expérience constructive n'est que partielle, étant donné que son utilisation n'est pas encore éprouvée et que sa structure reste expérimentale, elle a tout de même démontré la capacité de réinventer des assemblages structuraux, en réutilisant des matériaux en réemploi.

6. Vers une conception du project en matériau de réemploi

Au terme de cette recherche et du pavillon "Coup de poing" qui a été réalisé pour l'évènement Archifolies, nous avons identifié que l'environnement urbain est un réservoir important pour récupérer, sauvegarder, stocker et réutiliser les matériaux et les ressources qui constituent les bâtiments et notre cadre de vie.

D'autre part, en ayant un contact direct avec la matière et son enjeu constructif, la proposition architecturale émerge de la manipulation physique de matériaux, combinée avec des expérimentations par IA génératives de type Midjourney et Dall.E permettant d'étendre les expérimentations créatives.

La méthodologie de conception augmentée par l'aide de modèles de conception générative et par l'interaction avec un ou plusieurs concepteurs permet d'envisager un modèle prédictif de conception. Cette collaboration de type "cybernétique" présage de modèles créatifs futurs intéressants basés sur une hybridation entre modèles génératifs et nos compétences sensibles et cognitives. Cette évolution du fonctionnement de capacités est identifié par un neurologue tel que Raphaël Gaillard[5], comme une hybridation nouvelle, telle que celle engendrée par la lecture et l'écriture, et doit nous faire réfléchir sur un modèle pédagogique alternatif à l'apprentissage traditionnel du projet d'architecture.

1 Source: enquête "Déchets et déblais produits par l'activité de construction en 2014", SOeS

2 La démarche Zéro Artificialisation Nette (ZAN) est un objectif fixé pour 2050. Consacrée en 2018 par le Plan Biodiversité, puis en 2020 par la Convention citoyenne sur le climat, cette démarche consiste à réduire au maximum l'extension des villes en limitant les constructions sur des espaces naturels ou agricoles et en compensant l'urbanisation par une plus grande place accordée à la nature dans la ville.

3 Morin (Edgar) *La méthode* - Tome 3. La connaissance de la connaissance, Seuil, 1981, 416 pp.

4 Simon (Herbert A.) *The sciences of the Artificial*, 3rd Edition, The MIT Press Paperback, ISBN: 9780262691918, 1996, 248 pp

5 Gaillard (Raphaël). *L'homme augmenté, Futurs de nos cerveaux.* Grasset, 2024, EAN: 9782246835172, 352 pp.

RÉALITÉ AUGMENTÉE: VERS DES ESPACES IMMERSIFS INCLUSIFS

Jeux vidéo, espaces immersifs, métaverses... La multiplication des environnements numériques et l'élargissement de l'offre de services et d'activités qui leur sont associés augmentent progressivement le temps que nous passons immergés dans des espaces virtuels. Ces environnements rivalisent désormais avec notre réalité première, et parfois, notre perception brouille la distinction entre les deux, rendant l'expérience virtuelle indissociable du réel. Alors que nous passons plus de temps dans des espaces numériques immersifs que dans la réalité physique, qu'adviendra-t-il de notre relation avec notre propre corps?

Pour explorer ce phénomène, nous devons d'abord examiner les origines des espaces immersifs et les défis qu'ils présentent. Nous proposons ensuite une approche alternative de leur conception, qui ne les conçoit pas comme des domaines fictifs isolés, mais plutôt comme une extension de notre réalité primaire, une dimension augmentée qui s'intègre de manière transparente à notre monde physique.

Cette perspective nous permet d'envisager des points d'accès identifiables entre les espaces numériques et physiques, favorisant l'interaction et la continuité entre les deux domaines. Il est devenu crucial d'intégrer de tels portails dans la conception d'environnements immersifs, en veillant à ce qu'ils restent connectés à nos espaces quotidiens, qu'il s'agisse de maisons privées ou d'environnements urbains publics.

La fusion croissante des technologies et des services numériques dans les environnements urbains est en train de remodeler fondamentalement la façon dont les villes sont structurées et vécues. De la modélisation numérique des espaces urbains au traitement dynamique des données qui les redéfinit en permanence, la réalité mixte introduit une nouvelle forme d'interaction entre les mondes numérique et physique (Lucas, 2012). Elle permet à des individus géographiquement éloignés de collaborer et d'influencer collectivement la construction d'une réalité partagée. En s'appuyant à la fois sur l'héritage matériel et numérique, la réalité mixte nous oblige à repenser les processus d'hybridation, transformant ainsi notre perception des environnements urbains. Dans ce paradigme redéfini, la ville retrouve son essence fondamentale: un espace façonné par les interactions humaines..

1. La déconnexion entre les réalités virtuelle et physique

Malgré les progrès technologiques, la grande majorité des environnements immersifs restent des mondes parallèles détachés, n'offrant aucun lien tangible avec le monde physique. Lorsque nous enfilons un casque de réalité virtuelle ou un autre dispositif d'accès, nous ne passons pas à une réalité augmentée, mais nous faisons l'expérience d'une disjonction abrupte, entrant dans une dimension entièrement distincte. Il n'y a pas de seuils, pas de ponts, pas de chemins fluides facilitant le mouvement entre ces espaces. Au contraire, ces environnements imitent souvent les paysages des jeux vidéo ou les univers cinématographiques, dépourvus de continuité spatiale avec le monde réel.

Pour les architectes, cela représente un risque profond: un avenir dans lequel les individus mènent des vies fragmentées, oscillant entre une existence virtuelle de plus en plus déconnectée et un engagement décroissant avec l'espace physique. Une telle réalité pose un défi existentiel à l'architecture elle-même: si la conception spatiale devient purement numérique, que restera-t-il du rôle de l'architecte dans la conception de l'expérience vécue?

2. Repenser les environnements immersifs: augmenter la réalité, pas la remplacer

Notre analyse et notre expertise architecturale nous amènent à préconiser une approche différente: la conception d'environnements de réalité augmentée qui utilisent l'espace physique comme cadre de référence pour l'expansion virtuelle. Dans ce modèle, la réalité sert de point d'ancrage, permettant aux individus d'interagir avec des couches virtuelles tout en restant ancrés dans des références spatiales familières. Cette stratégie garantit que les expériences immersives n'existent pas de manière isolée, mais s'intègrent parfaitement au monde réel (Huggett, 2020).

Il existe un fossé persistant entre le patrimoine numérisé et la réalité virtuelle, ce qui soulève des préoccupations concernant la présence et le réalisme dans les environnements immersifs. Le contraste entre l'authenticité historique et les espaces simulés numériquement crée un paradoxe fondamental. Une solution proposée est le "réalisme libre", une approche qui cherche à combler ce fossé en alignant les reconstructions numériques sur l'expérience vécue. Le défi consiste à reconnaître la réalité virtuelle non seulement comme une simulation, mais comme un nouveau paradigme des systèmes d'information et des métavers, un cadre à travers lequel le patrimoine numérisé peut être pleinement intégré dans l'exploration et l'interprétation historiques.

Cependant, des études sociologiques et médicales récentes (Bojic, 2022) indiquent qu'une immersion prolongée dans des environnements virtuels entraîne des addictions et des dissonances cognitives importantes, qui font perdre aux utilisateurs leur conscience spatiale et corporelle. Le métaverse n'est donc pas seulement une plateforme d'expériences numériques, mais aussi un générateur d'insécurité spatiale et de dépendance psychologique.

3. L'émergence de mondes imaginaires et parallèles

La capacité humaine à conceptualiser et à habiter mentalement des espaces fictifs, mystiques et imaginaires remonte à la préhistoire. Les peintures rupestres de Chauvet, vieilles de plus de 60 000 ans, illustrent les premières formes de narration et de visualisation immersive. La grotte elle-même, isolée du monde extérieur, fonctionnait comme un portail vers un espace où des images s'animaient à la lueur vacillante des torches, transportant les premiers humains dans des royaumes spirituels, historiques ou aspirationnels.

Les environnements immersifs modernes, en particulier les métavers, présentent des similitudes frappantes avec ces pratiques ancestrales. Grâce aux progrès des casques de réalité virtuelle, des réseaux à haut débit et du rendu 3D en temps réel, les espaces numériques d'aujourd'hui offrent des niveaux d'immersion extraordinaires, qui font souvent paraître le monde réel terne en comparaison.

Le terme "métaverse", inventé par Neal Stephenson dans Snow Crash (1992), décrit une réalité secondaire, libérée des contraintes de la physique et de la matérialité. Ce monde fonctionne grâce à des portails, des centres de téléportation, des dispositifs de réalité virtuelle et même des améliorations biochimiques, un éventail de points d'accès qui brouillent la frontière entre le physique et le numérique. Les technologies émergentes telles que la blockchain, les NFT et les jumeaux numériques renforcent cette tendance, créant une économie virtuelle qui existe parallèlement au monde réel.

Pourtant, cette immersion est une arme à double tranchant. Elle libère l'esprit, mais le piège. Tout comme le sucre, le tabac, les réseaux sociaux et les plateformes de streaming, l'immersion numérique favorise la dépendance, monopolisant l'espace cognitif et émotionnel.

4. Le rôle de l'architecture à l'ère virtuelle

Pour les architectes, le métaverse représente à la fois un défi et une opportunité. L'absence de contraintes physiques (gravité, résistance des matériaux, limitations structurelles) permet une liberté de création sans précédent. Depuis les années 1990, les architectes et les artistes numériques ont exploré ces possibilités, des morphologies numériques de Greg Lynn aux théories urbaines computationnelles de Christopher Alexander. L'émergence de studios de design sans papier à l'université de Columbia et au Media Lab du MIT a encore démontré le rôle de l'architecture dans la conception d'environnements virtuels.

Cependant, l'architecture risque aujourd'hui de devenir obsolète dans le métaverse. Contrairement à la conception physique, où l'expérience spatiale est tangible et sensorielle, les environnements virtuels manquent souvent d'un langage architectural plus profond. De nombreux espaces de type jeu vidéo (Fortnite, Warfare) sont visuellement attrayants mais superficiels sur le plan architectural, reflétant une culture plus influencée par les romans graphiques et le cinéma que par la pensée architecturale.

Une préoccupation majeure est le coût environnemental de ces univers numériques. Le rendu en temps réel de mondes 3D immersifs exige une énorme puissance de calcul, ce qui contribue à la consommation d'énergie à l'échelle mondiale. D'ici 2024, les centres de données dédiés aux plateformes immersives devraient représenter 15% de la consommation mondiale d'électricité. Ne devrions-nous pas, par conséquent, rechercher des alternatives plus durables, où l'augmentation numérique complète plutôt que remplace l'espace physique?

5. Réappropriation de l'espace: vers un modèle hybride

En réponse à ces défis, nous proposons une vision alternative: la création d'environnements hybrides de réalité augmentée. Ces espaces intègrent des éléments physiques et numériques, garantissant que les expériences immersives restent ancrées dans le monde réel plutôt que de le supplanter.

Grâce à la recherche et à l'expérimentation, nous avons développé des modèles de réalité mixte au sein de la chaire Architecture & Intelligence et du laboratoire de réalité virtuelle de l'Université Paris Cité. Inspirés par l'approche atmosphérique de Peter Zumthor, nous mettons l'accent sur l'engagement sensoriel, en veillant à ce que les expériences immersives améliorent plutôt que remplacent la perception spatiale.

En établissant des portails du monde réel vers les univers numériques, nous empêchons le détachement total des individus de leur environnement physique. Nous devons résister à la dissolution de l'espace et plaider pour un avenir où l'architecture reste tangible, interactive et centrée sur l'homme.

Devons-nous nous résigner à des réalités parallèles ou investir dans des expériences spatiales plus significatives, multisensorielles et intelligentes? Cette réponse définit l'avenir de l'architecture et sa relation à notre corporalité.

À PROPOS DE L'AUTEUR

David Serero est architecte et urbaniste, connu pour ses recherches pionnières dans les domaines de l'innovation matérielle, de la géométrie et de la conception architecturale durable. Né à Grenoble, en France, il est diplômé de l'université Columbia de New York et de l'École d'architecture Paris-Villemin. En 2004, il reçoit le prestigieux Prix de Rome, une distinction qui a considérablement influencé sa carrière professionnelle. Après plusieurs années passées à New York, il déplace son agence en France, à Paris en 2006.

Le travail de David Serero se caractérise par une exploration distincte des matériaux et des géométries structurelles, avec un accent particulier sur les revêtements architecturaux. Sa conception de la Médiathèque de Bayeux illustre cette approche, intégrant un système structurel poreux qui optimise la diffusion de la lumière naturelle pour les espaces de lecture et les expériences multimédias.

Ses projets architecturaux ont été largement exposés en France et à l'étranger, dans des institutions prestigieuses telles que le Frac d'Orléans, le Pavillon de l'Arsenal, le MoMA de New York, l'Architectural League de New York, la Villa Médicis à Rome et le Mori Museum de Tokyo. Parallèlement à sa pratique architecturale, David Serero est professeur associé à l'École d'Architecture Paris-Val de Seine, où il enseigne et mène des recherches. Sa thèse de doctorat (2020) explore le potentiel inexploité des environnements urbains pour la production d'énergie éolienne, en exploitant les bords des bâtiments et les surfaces des façades comme éléments actifs de production d'énergie.

Serero est également cofondateur et codirecteur de la chaire de recherche Architecture & Intelligences, ainsi que co-directeur de l'EVCAU, un laboratoire de recherche de l'Université Paris Cité, qui étudie comment les ressources matérielles et les données numériques transforment la pratique architecturale.

Tout au long de sa carrière, il a occupé des postes d'enseignant au Pratt Institute de New York, à l'Université Columbia, à l'Université Paris Sciences et Lettres et à l'Université Paris Cité, tout en dirigeant de nombreux ateliers d'architecture et de recherche en France, en Italie, en Grèce, en Espagne, en Autriche et aux États-Unis.

Data
Energy
Matter

Architecture
Cimate
Changes

Published by
Actar Publishers,
New York, Barcelona
www.actar.com

Author
David Serero

Edited by
David Serero
Serero Architects
Paris, France
www.serero.com

Graphic design
Antoine Loyer
Marga Gibert

Copy editing
Actar / Serero Architects

Printing and binding
Arlequin SL

Distribution
Actar D, Inc. New York, Barcelona.

New York
440 Park Avenue South, 17th Floor
New York, NY 10016, USA
salesnewyork@actar-d.com

Barcelona
Roca i Batlle 2
08023 Barcelona, Spain
eurosales@actar-d.com

Indexing
English ISBN: 978-1-63840-048-6
Library of Congress Control Number: 2022943504

Printed in Spain

Publication date: November 2025